i

ISBN: 978-1-914934-95-7

A catalogue record for this book is available from the British Library

Published 2025 by
Northern Bee Books
Scout Bottom Farm
Mytholmroyd
Hebden Bridge HX7 5JS (UK)
01422 882751

www.northernbeebooks.co.uk

Book design, photography and photomicrographs by Lesley Jacques

Pollen photomicrographs taken using Olympus TG-6 mounted on Brunel SP100 trinocular compound microscope

Electron micrographs by Lesley Jacques with thanks to the Natural History Museum, London

Where do the bees go?

Lesley Jacques, NDB

For dear friends, Christine Coulsting, Mary Hunter,
Jane Medwell, Rhona Toft and Helen Tworkowski,
for support and truly excellent discussions, and without
whom this would have been a lonely pursuit indeed.

George Clifford's Herbarium Sheet

*From the collection of George Clifford that helped to lead
Carl Linnaeus to establish the binomial nomenclature of all
living things.*

Natural History Museum, London, 14 February 2024.

Preface

As a nation of bee lovers, we have an increasing awareness of the importance of not only bees, but pollinators of all kinds in our environment and ecosystem. Pollinator ecology is intimately tied to the floral sources that provide key nutrients for insect species, so it follows that any study of insects will eventually come around to their foraging habits and relationship with plants.

This work was conceived as part of a study programme for the UK National Diploma in Beekeeping. As such it provides information which is focused on UK floral species that are the most important forage species for the honey bee, *Apis mellifera*. However many of these plants are equally valuable to other insect pollinators, notably our wide variety of other bee species, hover flies, moths, butterflies and beetles.

Historically, botanical studies have relied upon pressed samples—herbaria—to provide reference material and a permanent record for scientific study. The earliest herbaria date back to the sixteenth century; these beautiful but fragile documents are generally held in national archives under controlled conditions (in the UK, there are notable collections at the Natural History Museum and the Linnean Society, both in London). Some are available to view on request. A pressed sample has a number of advantages over a photograph, not least of which is the level of detail that is recorded, not to mention the preservation of genetic material. However the two serve different purposes, and one cannot replace the other. With that in mind, the original work from which this is drawn centred around the collection and preservation of pressed samples, augmented with a photographic record of each species, some taken at the time of collection and some gathered subsequently to capture different life stages of the plant. Scans of the pressed samples form the primary entry for each species. Each contains general information about the plant family, flowering periods, habit and associated benefits to insects. Photomicrographs are included of stained and mounted pollen grains from each species.

I cannot pretend that this is a complete record; that would be more than a lifetime's endeavour. But it does include many of those species that will be of interest to beekeepers, hopefully to budding entomologists, and possibly to those with an ongoing curiosity around the pollinators in their gardens.

Lesley Jacques NDB, January 2025.

Contents

Contents

Content notes

Data in this volume are referenced appropriately where a credible source exists. Not all data are available consistently for all species, so there is necessarily some variability in the parameters reported here.

The number of accepted species by family is being updated constantly. Therefore the figures that are given here should be checked before citing them with confidence.

Bee forage potential

Data are given as reported by Kirk and Howes for different bee types; honey bees, short-/long-tongued bumble bees or solitary bees.[1] Those plant species yielding pollen are indicated by P or p for major/minor source. Likewise nectar sources by N or n.

Notes on pollen data

The nutritive value of pollen can be expressed in terms of its percentage crude pollen content. A range of 20-25% is considered a minimum to be adequate to meet the protein requirements of a honey bee colony.[2] Examples covered in this work which exceed this include *Brassica napus* (oil seed rape), *Onchobrychis viciifolia* (sainfoin) and *Echium vulgare* (Viper's bugloss), among others. Also other species outside of the current scope of this work include *Vicia* spp. (vetch), and *Cirsium vulgare* (spear thistle)—all ubiquitous plants which are important forage species. In some cases, the nutritional value of pollens to insects can be inferred. Data around crude protein percentages are available for pollen of some species and are cited where possible—those falling into the range >25% can generally be considered to provide the most nutritional value to foragers, with the high percentage protein helping to mitigate any imbalance in the amino acid profile of the pollen.[2]

Some pollens are particularly high in essential amino acids (those amino acids which cannot be synthesized by the body, but are required for normal and proper physiological functioning, and therefore must be consumed in the diet). For honey bees, essential amino acids include arginine, isoleucine, leucine, methionine, phenylalanine, threonine and valine.

Pollen coefficients[3]

A single source honey extracted from pollen-free combs will contain a specific proportion of pollen; this value is fixed for each type of honey and is known as the pollen coefficient. It is highly dependent on the structure and form of the flower and position of the nectaries. Generally, if the anthers are directly above the nectaries, then it follows that pollen can fall into the nectar, and more pollen would be found in honey which is derived from that plant. If the anthers hang below the nectaries, then the converse is true.

The majority of species have a pollen coefficient of 20-80; the quantity of pollen in a monofloral honey is as one would expect and is said to be normally represented. A pollen coefficient 1-20 is found in some species; of note here are some *Lamiaceae* family species (thyme, rosemary, mint), *Erica* spp. (heaths and bell heather), *Helianthus annuus* (sunflower), *Tilia* (lime) *Taraxacum* spp. (dandelion) and *Calluna vulgaris* (ling heather). In these cases, there is less of the pollen than would be expected in the honey and it is said to be under-represented.

Conversely, pollen is greatly over-represented (pollen coefficient >100) in e.g. *Brassica napus* (oil seed rape), *Echium vulgare* (Viper's bugloss), *Myosotis* spp. (forget-me-not) and others.

Pollen aperture morphology

A number of terms used to describe pollen morphology warrant further explanation, in particular, those terms which describe the aperture formation. The following terms are used throughout:[4]

Colpate—with colpi (furrows)
Porate—with pores
Colporate—with pores and colpi
Sulcate—with a groove running along its length

Honey potential

In her work *Honey—A Comprehensive Survey*,[5] Eva Crane cites the 'honey potential' of 200 flowering plants that are worked by honey bees. This represents the maximum quantity of honey that could be obtained in the course of a season from 1 hectare of land sown with the species in question. Optimal growing conditions are assumed, and an adequate foraging force to collect all nectar secreted. This is very much a theoretical value, but does give a good measure of the relative yield of some of the species described here. Her classifications were as follows:

Class 1—0-25 kg/Ha   Class 4—101-200 kg/Ha
Class 2—26-50 kg/Ha   Class 5—201-500 kg/Ha
Class 3—51-100 kg/Ha   Class 6—>500 kg/Ha

Papaveraceae—Poppy

Papaveraceae, the poppy family of flowering plants includes 42 genera, with 775 species,[6] 50 of which are of the genus Papaver (poppy). Most in this family are herbaceous plants though there are a few woody shrubs. Flowers are bisexual and generally dish-shaped with 4-6 petals, one superior pistil and many stamens. The fruit is a spherical or linear capsule. The leaves are usually deeply cut or divided into leaflets.[7,8]

Family	Papaveraceae
Species	*Papaveraceae* spp., *Papaver rhoeas*
Common name	Common poppy, corn poppy, Flanders poppy
Flowering time	June-September

Species notes

Papaver rhoeas, **common poppy,** is a variable erect annual which can reach ~70 cm in height. Flowering is generally in the late spring although more flowers can appear in the early autumn under favourable conditions.

Stems bear a single showy flower, 5-10 cm across, most usually red, with slightly overlapping petals. The flower stem is usually covered with coarse hairs that are held at right angles to the surface making the plant feel slightly bristly. The capsules are hairless, ovoid (egg-shaped), less than twice as tall as they are wide, with a stigma at least as wide as the capsule. Seeds are dispersed from holes at the top of the ripe capsule when the plant is moved by the wind. Like many other species of *Papaver*, the common poppy exudes sticky white sap when the tissues are broken.

Cultivars of various colours are not uncommon, usually white through pink shades.

Habitat

Common poppy is widely distributed, mostly across the northern hemisphere, preferring scrubby or disturbed, well drained ground and grassland. Seeds can lie dormant in the ground for a long time.

Associated insect benefits

There is no notable nectar presentation. Poppy is primarily a pollen source which produces abundant dark blue-grey pollen which attracts numerous pollinators including honey bees.[9] Although the crude pollen content is not classed as high (21.4%), 36% by weight comprises essential amino acids, making poppy pollen of particular nutritional value.[2]

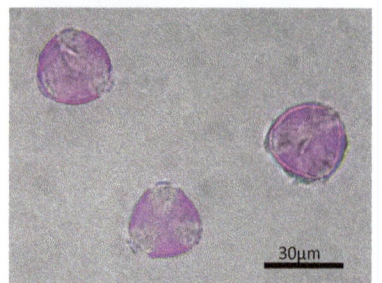

Bee forage potential[1]		Notable foraging bee species[1]
Nectar/pollen	P	Honey bee, *Apis mellifera*
Honey bees	✿ ✿	Mining bees, *Halictus* spp.
Short-tongued bumble bees	✿ ✿	Poppy mason bee, *Hoplitis papaveris* (in
Long-tongued bumble bees	✿	mainland Europe)
Solitary bees	✿ ✿	

Pollen presentation

Pollen from this species is a characteristic blue-black colour.[10,11] The grains are ~25-30 μm in diameter with three colpate apertures. Light microscopy shows a rounded triangular cross section in polar view. The pollen grains have a thin exine with a granular surface structure.[12]

30μm

Papaveraceae—Poppy

Ranunculaceae—Wild Clematis

Ranunculaceae is a large plant family comprising over 2200 species in 62 genera.[1] Most are flowering plants, widely distributed in all temperate and subtropical regions. The leaves are usually alternate and stalkless and may be simple or much divided, often with sheathing bases. The flowers usually have two to five free sepals and may be radially symmetrical or irregular.

Family	Ranunculaceae (Buttercup family)
Species	*Clematis vitalba*
Common name	Wild clematis, traveller's joy, old man's beard
Flowering time	July-August

Species notes

Clematis vitalba, wild clematis is a vigorous climbing shrub with branched, grooved stems, deciduous leaves, and scented greeny-white flowers with fluffy underlying sepals (B). The many fruits formed in each inflorescence have long silky appendages which give the characteristic appearance of fluffy whiskers, gaining the plant one of its common names of 'old man's beard'. Seeds are easily dispersed by many means; C. vitalba can also reproduce vegetatively by regrowing from roots and stems and its vigour has resulted in its classification as an invasive species in some territories, notably New Zealand and Canada. The plant can reach a height of 7m plus, supporting itself with twining leaves and tendrils which wrap around and grasp anything in their path. Leaves are pinnately compound, composed of three to five leaflets (C).[9,13]

Habitat

C. vitalba prefers base-rich alkaline soils and a moist climate with warm summers. It is commonest in southern England (south of a line between the Mersey and Humber estuaries), although will grow as a planted species elsewhere.

Associated insect benefits

There is no notable honey crop, however flowers do provide forage for a variety of bees including honey bees and both short and long-tongued bumble bees. Nectar presentation is unusual, with nectar being produced on the filaments rather than from nectaries.[1,9]

This species is also eaten by the larvae of a wide range of moths. This includes many species which are reliant on it as their sole food plant; including small emerald (*Hemistola chrysoprasaria*), small waved umber (*Horisme vitalbata*) and Haworth's pug (*Eupithecia haworthiata*).[13]

Bee forage potential[1]		Notable foraging bee species[1]
Nectar/pollen	N	Honey bee, *Apis mellifera*
Honey bees	✿ ✿	Buff-tailed bumble bee, *Bombus terrestris*
Short-tongued bumble bees	✿ ✿	White-tailed bumble bee, *Bombus leucorum*
Long-tongued bumble bees	✿	
Solitary bees	✿	

Pollen presentation

Pollen loads are grey-white in colour.[11] The pollen grains are ~20 μm in diameter. Light microscopy shows a rounded triangular cross section in polar view and three colpate apertures. The pollen grains have an exine of medium thickness with a granular surface structure.[10,14]

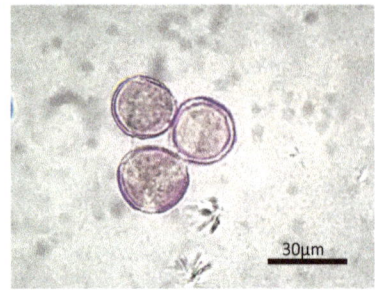

30μm

Grossulariaceae—Currants

The Grossulariaceae family includes a single genus, *Ribes,* which includes ~150 species falling into two distinct groups; currants and gooseberries.[6] Most are native to the temperate regions of the northern hemisphere. Many are widely cultivated either for edible gooseberry or currants (blackcurrant, redcurrant, white currant) or as ornamental cultivars (e.g. the flowering currant *Ribes sanguineum*).

Family	Grossulariaceae
Species	*Ribes nigrum / Ribes rubra*
Common name	Black currant, red currant
Flowering time	March-April

Species notes

Ribes nigrum/Ribes rubra, **black currant** and **red currant** are medium-sized shrubs, growing up to 1.5 metres. The leaves are 5-lobed and alternate, with a serrated margin (A). In both species, flowers are produced in racemes known as 'strigs' (B) up to 8 cm long and containing up to 20 flowers, each about 8 millimetres (3⁄8 in) in diameter. *R. rubra* flowers are generally whitish-green; *R. rubra* flowers show a pink tinge. Each flower has a hairy 5-lobed calyx with yellow glands, the five lobes of which are longer than the inconspicuous petals. There are five stamens surrounding the stigma and style and two fused carpels. The flowers open in succession from the base of the strig and are mostly insect pollinated, but some pollen is distributed by the wind.[9,15]

Habitat

Currants can grow well on sandy or heavy loams, or forest soils, as long as their nutrient requirements are met. They prefer damp, fertile but not waterlogged ground and are intolerant of prolonged dry conditions.[9,15]

Associated insect benefits

While currant is not generally assumed to be a major honey-yielding species, Crane lists red currant specifically as potentially producing 101-200 kg/Ha under optimum conditions.[5] The flowers produce a sucrose-dominant nectar[2] and are visited by a variety of pollinators including short and long-tongued bumble bees, honey bees and solitary bees.[1]

Bee forage potential[1]		Notable foraging bee species[1]
Nectar/pollen	N/p	Honey bee, *Apis mellifera*
Honey bees	✿✿✿	Bumble bee species generally, notable
Short-tongued bumble bees	✿✿✿	pollen forage by buzz pollination
Long-tongued bumble bees	✿✿	
Solitary bees	✿✿	

Pollen presentation

Pollen loads of both *R. nigrum* and *R. rubra* are yellow.[10] The pollen grains are rounded, ~30 μm in diameter with multiple porate apertures.[16] The exine has a smooth surface and a medium thickness.[16] It would be difficult to differentiate the pollen grains of these two species with light microscopy. (*R. nigrum* pollen shown here.)

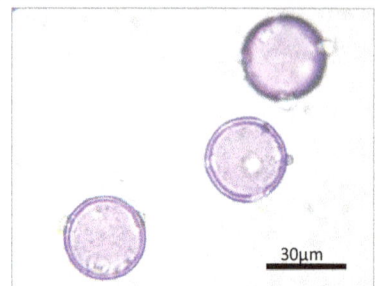

30μm

Grossulariaceae—Currants

Ribes rubra

Ribes sanguineum

Related ornamentals

The flowering currant, *Ribes sanguineum*, is a common ornamental across much of the UK. It is among the earliest flowering shrubs and beneficial to bees. Generally flowering in April, the flowers are accessible to both long and short-tongued species. Honey bees work the flowers, particularly for pollen. There are many varieties of flowering currant, some of which have blossoms with a long flower tube, only accessible by longer-tongued species.[1]

Grossulariaceae—Gooseberry

Family	Grossulariaceae
Species	*Ribes uva-crispa*
Common name	Gooseberry
Flowering time	April-May

Species notes

Ribes uva-crispa, **gooseberry** is a medium-sized spiny shrub, growing up to 1.5 metres. The leaves are 3 or 5-lobed and alternate, with a crenated margin (A). The small bell-shaped flowers are produced singly or in small clusters of and are usually greenish through pinkish depending on the variant (B,C).

Habitat

Gooseberry is native to the temperate regions of the northern hemisphere and is extremely hardy, being known to survive as far north as the Arctic circle. They thrive in moist heavy clay soil and cool humid climates.[17]

Associated insect benefits

Gooseberry is an early flowering species, and as such provides a valuable source of forage when there may be little available for insects.[9] Although this is not considered a particularly high-yielding species for honey bees, Crane (1979) suggests that under optimum conditions gooseberry may produce honey up to 51-100 kg/Ha although there are conflicting data regarding the value of the nectar, with different authors suggesting anything from 0.13 to 5.41 mg of sugar per flower per day (24h).[5] Gooseberry nectar is sucrose-dominant.[2]

Gooseberry is also host to a number of Lepidoptera, namely caterpillars of comma butterfly (*Polygonia c-album*),[18] magpie moth (*Abraxas grossulariata*), V-moth (*Macaria wauaria*)[6] as well as the gooseberry sawfly (*Nematus ribesii*).[19]

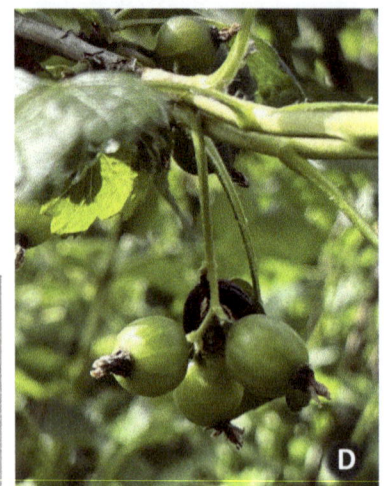

Bee forage potential[1]		Notable foraging bee species[1]
Nectar/pollen	N	Honey bee, *Apis mellifera*
Honey bees	⚜⚜	Early-year bumble bees and solitary bees
Short-tongued bumble bees	⚜	e.g. Tawny mining bee, *Andrena fulva*
Long-tongued bumble bees	⚜	
Solitary bees	⚜	

Pollen presentation

Pollen loads of *R. uva-crispa* are yellow.[10] The pollen grains are rounded ~30 μm in diameter with multiple porate apertures. The exine has a smooth surface and a medium thickness.[20] It can be difficult to differentiate the pollen grains of these currant species under light microscopy.

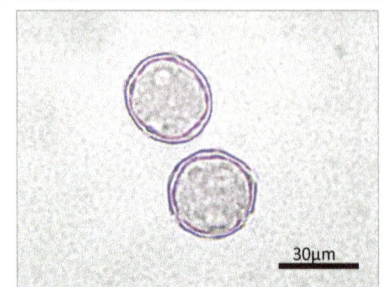

30μm

Grossulariaceae—Gooseberry

Fabaceae—Field Bean

The Fabaceae or commonly known as the legume, pea, or bean family, is a large and agriculturally important family of flowering plants which includes trees, shrubs, and perennial or annual herbaceous plants. The family is widely distributed, and is the third-largest land plant family in terms of number of species, behind only Orchidaceae and Asteraceae. There are some 765 genera and nearly 20,000 known species.[21]

Family	Fabaceae—pea family and legumes
Species	*Vicia faba*
Common name	Field bean, broad bean
Flowering time	May-June

Species notes

Vicia faba, **field bean** is erect slightly fleshy plant reaching up to 1.8 m in height. It typically has two-four stems with a characteristic square cross-section. The stem and branches are crowded with short-petioled glaucous compound leaves (A).

Bean flowers bear the characteristic form of their family, having 5 united petals that form a distinct flower shape. The flowers have as many as 10 stamens (sometimes fewer) and a partially inferior ovary with a single carpel (B,C).[21,22]

Habitat

Vicia faba is an annual plant widely cultivated as a food and cover crop (D), and although preferring rich loams, it is a tolerant species and will thrive in most soil conditions and environments including salinity. It is sufficiently frost-tolerant to survive winter (provided temperatures do not drop below 9-12°C). This hardiness means that in some areas, field bean can be sown in autumn ahead of maturation the following spring. It is, however, drought intolerant.

Associated insect benefits

Vicia faba is a major forage crop for a number of bee species. When worked by honey bees it yields rapidly-granulating honey with a light colour and mild flavour.[5] Nectaries are sited deep within the flower structure, enabling only long-tongued bees to access the nectar through the usual means. However close examination of the blooms will often reveal a hole bitten in the back of the flower where shorter-tongued species have gained access, termed secondary nectar robbing. Field beans also have extra-floral nectaries on the underside of the stipules which secrete in good weather.

Field bean is known to harbour aphids, specifically the black bean aphid (*Aphis fabea*), and so can be a source of honeydew.

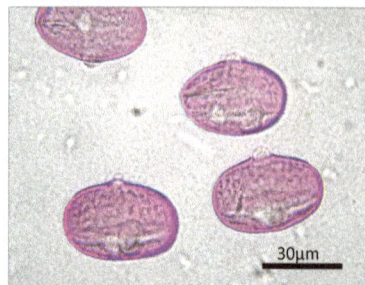

Bee forage potential[1]		Notable foraging bee species[1]
Nectar/pollen	N/P	Garden bumble bee, *Bombus hortorum*
Honey bees	🐝🐝🐝	Common carder bee, *Bombus pascuorum*
Short-tongued bumble bees	🐝🐝🐝	Ruderal bumble bee, *Bombus ruderatus*
Long-tongued bumble bees	🐝🐝🐝	Honey bee, *Apis mellifera* via secondary
Solitary bees	🐝🐝🐝	nectar robbing

Pollen presentation

Pollen loads of *Vicia faba* are pale grey-green.[10,11] The pollen grains are oval, measuring ~40x30 μm in diameter with three colporate apertures. The thin exine has a smooth surface.[23] The pollen grains show characteristic internal granularity when examined with the light microscope. This species has a pollen coefficient of 35 and is normally represented in honey.[3]

30μm

Fabaceae—Field Bean

Fabaceae—Red Clover

Clover is of the genus *Trifolium* (three-leaved) - a genus of some 300 annual and perennial species in the Fabaceae family. Clovers occur in most temperate and subtropical regions other than south-east Asia and Australia. They are important agriculturally, being planted as livestock feed, a cover crop or green manure. Flowers are highly attractive to bees and clover honey is an important product of clover cultivation.[24]

Family	Fabaceae
Species	*Trifolium pratense*
Common name	Red clover
Flowering time	May-September

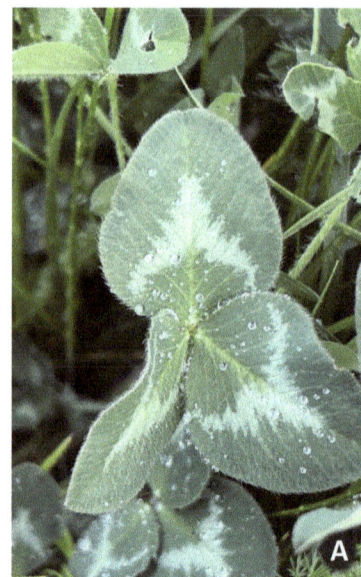

Species notes

Trifolium pratense, **red clover** is a short-lived herbaceous perennial or biennial. Variable in size, it can reach 60 cm in height though the straggling stems require support. The characteristic leaves are alternate, trifoliate (three leaflets), each leaflet being 15–30 mm by 8–15 mm; dark green with a pale crescent to the outer half of the leaf; the petiole is 1–4 cm long, with two basal stipules that are abruptly narrowed to a bristle-like point. Leaves can appear lightly downy (A). The flowers are mid-pink, rarely pale pink/white, with a paler base, with corollas some 12–15 mm long, in a dense inflorescence (B,C).[5, 25]

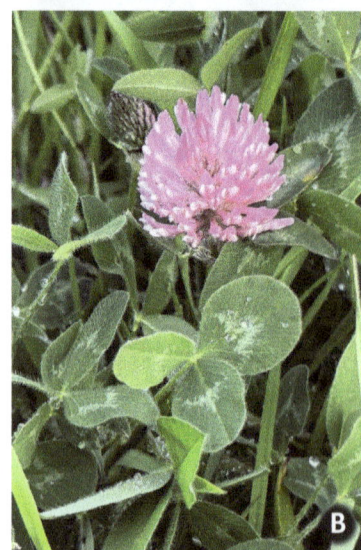

Habitat

Trifolium pratense is native to Europe, western Asia and northwest Africa, but has naturalised elsewhere. It is a hardy species and will grow abundantly in well-drained soils, preferring a sunny aspect. Red clover also has a deep tap root, which gives it good drought tolerance. It is grown as a fodder crop, and is valued for its nitrogen fixation, which increases soil fertility.

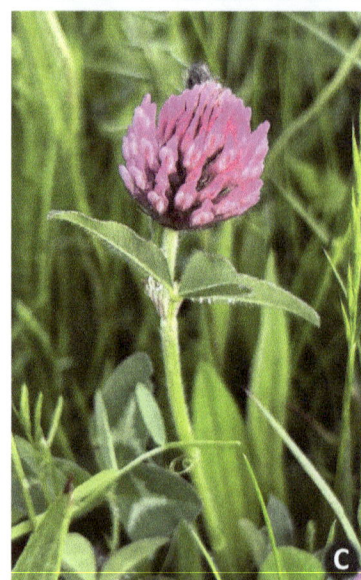

Associated insect benefits

The corolla of *Trifolium pratense* is sufficiently long that access to the nectary is restricted to longer tongued species—bumble bees, for which it is a major forage source. Under managed conditions, smaller flowers may result, with nectar that is accessible to shorter-tongued species. In favourable conditions, a build-up of nectar in the corolla will be accessible to honey bees. Under these circumstances, red clover may yield up to 500 kg/Ha of honey. Red Clover yields a pale, mild-flavoured honey which granulates rapidly.[5]

Honey bees are able to take pollen from red clover. The pollen is highly beneficial, with a crude protein content of 23-35%, 41% of which is comprised of essential amino acids.[2]

In both this species and *Trifolium repens* (white clover), the florets are held erect until pollination occurs, after which the nectaries cease to produce and the florets droop, thus signalling the absence of yield to pollinators.

Bee forage potential[1]		Notable foraging bee species[1]
Nectar/pollen	N/P	Garden bumble bee, *Bombus hortorum*
Honey bees	✿ ✿	Common carder bee, *Bombus pascuorum*
Short-tongued bumble bees	✿ ✿	Great yellow bumble bee, *B. distinguendus*
Long-tongued bumble bees	✿ ✿ ✿	Shrill carder bee, *B. sylvarum*
Solitary bees	✿ ✿	Short-haired bumble bee, *B. subterraneus*

Pollen presentation

Pollen loads from *Trifolium pratense* are brown.[10,11] The round pollen grains are ~30 µm in diameter, with three colporate apertures. The exine is thin, and has a slightly netted (reticulate) surface texture.[26] This species has a pollen coefficient of 25 and is normally represented in honey.[3]

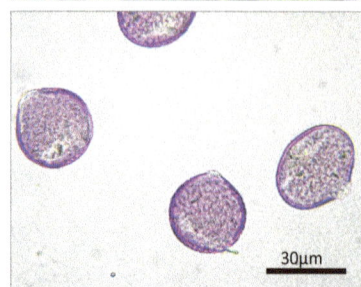

30µm

Fabaceae—White Clover

Family	Fabaceae
Species	*Trifolium repens*
Common name	White clover
Flowering time	May-September

Species notes

Trifolium repens, **white clover** is a low, creeping perennial or biennial. The main stems are prostrate, forming roots at the nodes. Leaves and flowers arise form these horizontal stems, growing upward. The plant can reach 40-50 cm in height though it is often low growing in grass which is grazed or mown. Like others in this family, the characteristic leaves are, trifoliate (with three leaflets), each leaflet being 10–30 mm long and very finely toothed, usually with a pale chevron near the base. Leaves are largely hairless (A). Flowers are held on leafless, upright stems. Each inflorescence is made up of a globular cluster of florets, each approximately 7-10mm in length, which are usually white, sometimes fading to pale pink as the flower ages (B,C).[1]

Clover is leguminous and therefore a valuable crop for soil improvement, adding about 55–170 kg per hectare (about 50–150 pounds per acre) of nitrogen to the soil and increasing the availability of other nutrients for following crops. As such, white clover was once widely cultivated as a fodder or cover crop.[2]

Habitat

Preferred habitat includes meadows, pasture and calcareous grasslands, and although the species is generally ubiquitous it will be less common in acid or heavy, wet soils.

Associated insect benefits

Clover is highly beneficial to insects, providing both nectar and pollen. The corolla of white clover is shorter than that of red clover, making the nectaries accessible to shorter tongued pollinator species such as honey bees and some solitary bees.[9] It is said that clover once accounted for up to 75% of the UK honey crop when it was used to fortify grazing pasture.[9]

Similarly to *Trifolium pratense* (red clover), the florets are held erect until pollination occurs, after which the nectaries cease to produce and the florets droop (D). White clover is a major honey crop in temperate regions, particularly Europe and New Zealand, providing a light, mild honey that is slow to granulate. A managed clover crop might produce 50-200 kg/Ha of honey.[5]

Clover pollen has a crude protein content of 22.5-35.4%[2] placing it among those forage species which offer greatest nutritional benefit to pollinators.

Bee forage potential[1]		Notable foraging bee species[1]
Nectar/pollen	N/P	Mining bees, *Lassioglossum* spp.
Honey bees	✿ ✿ ✿	Leafcutter bees, *Megachile* spp.
Short-tongued bumble bees	✿ ✿ ✿	Honey bees, *Apis mellifera*
Long-tongued bumble bees	✿ ✿ ✿	Bumble bees generally
Solitary bees	✿ ✿	

Pollen presentation

30μm

Pollen loads from *Trifolium repens* are brown.[10,11] The round pollen grains are ~30 μm in diameter with three colporate apertures. The exine is thin, and has a slightly netted (reticulate) surface texture.[27] This species has a pollen coefficient of 25 and is normally represented in honey.[3]

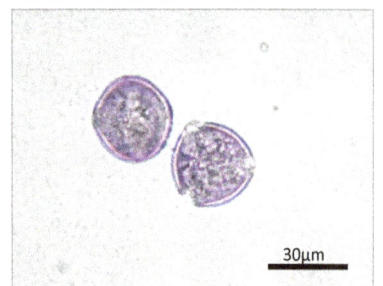

Fabaceae—Sainfoin

Family	Fabaceae
Species	*Onobrychis viciifolia*
Common name	Sainfoin
Flowering time	June-August

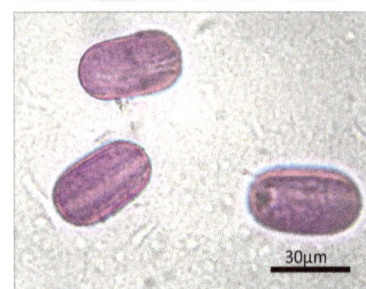

Species notes

***Onobrychis viciifolia*, sainfoin** is an erect, leggy and deep-rooted perennial bearing showy spikes of pink flowers from June through August. The plant is sparsely hairy, with leaved divided into 6-14 pairs of leaflets, each 10-35 mm long with a pointed tip and a single, terminal leaflet (A). Stipules are papery, small and pointed. Flowers vary from 10-14 mm and are typical of the Fabaceae form.[25] They comprise a compact raceme with 20-50 florets, each bright pink and delicately veined (C).

Sainfoin was introduced in the 17th century as a fodder crop, and was once widespread in the UK. It is little-used now, in spite of many agricultural benefits.[25] As well as being a nitrogen fixer, sainfoin has been found to increase the sequestration of other nutrients such as phosphates, into the soil, and is advocated by *The Soil Association* as an ideal rotation crop ahead of cereals or brassicas.[28] Its large and deep tap root confers good drought tolerance.

Habitat

Found mostly in the south of the UK, preferred habitat includes dry grassy places over calcareous soils.[25] It will not thrive in soils of pH>6.2 or in wet soils.[29]

Associated insect benefits

Sainfoin is one of the UKs most beneficial species for pollinators, by virtue of an extended flowering period and rich nectar yields. The flowers produce copious nectar and are particularly attractive to honey bees[30] as well as to a wide range of bumble bees, butterflies and other invertebrates. Sainfoin is said to produce more honey than any other species in the legume family,[28] although Crane does not rate it as highly as e.g. red clover, suggesting yields of 5-200 kg/Ha under optimum conditions.[5] Nectar secretion is reliable at relatively low temperatures (down to ~14°C).[30] The honey is pale/light amber, very sweet, and with a more pronounced flavour than that from other legumes.[5]

Sainfoin pollen has a crude protein content of up to 32.5% and is among those forage species which offer high nutritional benefit to pollinators.[2]

Bee forage potential[1]	
Nectar/pollen	N/P
Honey bees	🐝🐝🐝
Short-tongued bumble bees	🐝🐝🐝
Long-tongued bumble bees	🐝🐝🐝
Solitary bees	🐝🐝🐝

Notable foraging bee species[1]
Many *Bombus* species
Honey bee, *Apis mellifera* prefers sainfoin when available
The rare mining bee *Melitta dimidiate* is associated with sainfoin abundance

Pollen presentation

Pollen loads from *Onobrychis viciifolia* are brown.[10] The ovoid pollen grains are ~30-40 µm in length with three colpate apertures. The exine is thin, and has a slightly netted (reticulate) surface texture.[31] This species has a pollen coefficient of 75 and is normally represented in honey.[3]

30µm

Fabaceae—Bird's Foot Trefoil

Family	Fabaceae
Species	*Lotus corniculatus*
Common name	Bird's-foot trefoil, eggs and bacon
Flowering time	May-September

Species notes

Lotus corniculatus, **bird's-foot trefoil** is a ubiquitous hardy perennial legume forming a sprawling, low-growing and largely hairless plant. Leaves are divided into five leaflets' with the basal pair at the junction of leaf and stalk resembling stipules (the true stipules are tiny). The height of the plant is variable, from 5-20 cm, occasionally more where there is supporting undergrowth.

Flowers are bright yellow, sometimes streaked with red, particularly when in bud, and are held at the top of long stalks. Florets are clustered in groups of 2-8 (A,B).[25] This plant has many common names in the English language, most of which allude to its red/yellow colouring.

Seed pods are similarly in groups and their likeness to a bird's foot is what gives this plant its most common name (C).[25]

Habitat

Found across the UK, preferred habitat includes dry grassy places. It is common in unimproved grassland as well as shingle banks and dunes.[25]

Associated insect benefits

Like sainfoin, Bird's-foot trefoil has a deep tap root, rendering it relatively insensitive to moisture. As such, it offers a consistent nectar source that is attractive to a range of pollinators including short and long-tongued bumble bees, and solitary bees.[9] It is mostly visited by bumble bees. It is also an important forage species for the common blue butterfly (*Polyommatus icarus*) and an important food plant for a number of other Lepidoptera species including six-spot burnet moth (*Zygaena filipendulae*) and silver-studded blue butterfly (*Plebejus argus*) .[32] This species has also been shown to be highly attractive to a number of ubiquitous hoverfly species, namely the long hoverfly (*Sphaerophoria scripta*), marmalade hoverfly (*Episyrphus balteatus*) and hairy-eyed flower fly (*Syrphus torvus*).[33]

Although honey yield is low, this species is said to produce a light honey of good quality, which will granulate rapidly. Crane suggests a very variable honey yield from 20-200 kg/Ha, which may be due the influence of environmental conditions.[5]

Bee forage potential[1]		Notable foraging bee species[1]
Nectar/pollen	N/P	Red tailed bumble bee, *Bombus lapidarius*
Honey bees	🐝🐝	Common carder bee, *Bombus pascuorum*
Short-tongued bumble bees	🐝🐝🐝	*Osmia inermis, Osmia uncinata,*
Long-tongued bumble bees	🐝🐝🐝	*Osmia xanthomelana* (all rare)
Solitary bees	🐝🐝🐝	*Megachile* and *Colletes* spp.

Pollen presentation

Pollen loads from *Lotus corniculatus* are brown.[10] The ovoid pollen grains are ~20 μm in length with three colporate apertures. The exine is thin with minimal surface texture.[10,34] This species has a pollen coefficient of 25 and is normally represented in honey.[3]

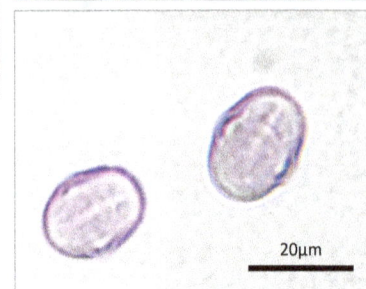

Rosaceae—Hawthorn

The Rosaceae family includes 4,828 known species in 91 genera distributed worldwide.[9] Most are annual or perennial shrubs, trees or herbs, mostly deciduous but with some evergreen species occurring. Most species have alternate leaves which can be simple, trifoliate, palmate or pinnate. Flowers are generally bisexual and typically have 4-5 sepals and a similar number of petals with at least 5 stamens, often many more.[35]

Family	Rosacea
Species	*Crataegus monogyna*
Common name	Hawthorn, thorn-apple, May blossom, hawberry
Flowering time	April-June

Species notes

Crataegus monogyna, **hawthorn** is an abundant deciduous hedgerow and woodland species forming a large thorny shrub or small tree up to 10-15 metres.[25]

The plant has sparsely hairy to hairless leaves 15-50 mm long, deeply cut into 3-7 pointed lobes (number of lobes depend on the maturity of the branch) (A).

White flowers are borne in flat-topped clusters from late April to June, giving the plant its common name of 'May blossom'. Each has 5 petals some 4-6mm long, and a cluster of white or sometimes pink anthers (B,C). In the autumn, reddish-brown fruit (haws) are borne in clusters (D). These form an important winter foodstuff for birds, particularly thrushes and waxwings.

The closely related Midland hawthorn, *Crataegus laevigata*, found mainly in the south of England, is less stiff, with more shallowly lobed leaves and an earlier flowering period. The two species frequently hybridise, producing fully fertile offspring.[25]

Habitat

Hawthorn is ubiquitous throughout the British Isles, and will flourish in all soil types, with the exception of acid peat.

Associated insect benefits

Hawthorn flowers plentifully and the shallow flowers are accessible to a wide variety of insects, including honey bees and both long— and short-tongued bumble bees and many flies.[36] Although it is common to see honey bees working the hawthorn, they generally do so for pollen. Nectar yield is uncommon and said to require temperatures consistently above 25°C (rare in much of the UK in May) to let nectar in any quantity. On those rare occasions when a honey crop is possible, it results in a 'nutty' flavoured, amber honey.[5] Hawthorn is used as a food plant by many invertebrates, including the aptly named true bug, the hawthorn shield bug (*Acanthosoma haemorrhoidale*) and the larvae of a number of Lepidoptera species, including the small eggar moth (*Eriogaster lanestris*).

Bee forage potential[1]		Notable foraging bee species[1]
Nectar/pollen	N/P	Open flowers are accessible to all short-tongued species
Honey bees	✿ ✿ ✿	
Short-tongued bumble bees	✿	Honey bee, *Apis mellifera*
Long-tongued bumble bees	✿	*Andrena, Halictus* and *Lasioglossum* spp.
Solitary bees	✿ ✿ ✿	

Pollen presentation

Pollen loads of *Crataegus monogyna* are grey-green/yellow[11], and pollen grains ~35-50 µm, oval-triangular in cross section with three colporate apertures.[37] The surface structure is smooth with few obvious surface features.

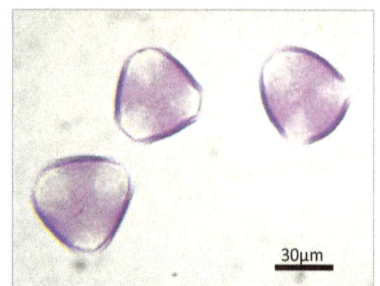

30µm

Rosaceae—Cherry

Family	Rosaceae
Species	*Prunus* spp. (Variety shown is *Prunus avium)*
Common name	Cherry, stone fruit
Flowering time	March-April

Species notes

Prunus is a substantial genus of trees and shrubs including plums, cherries, peaches, and almonds, many of which are cultivated widely for either fruit of as ornamentals. Generally, leaves are simple and lanceolate, sometimes finely toothed (A), often with extra-floral nectaries on the leaf stalks. Most bear conspicuous white to pink flowers (sometimes deep pink), typically with five petals, five sepals and numerous stamens. Flowers emerge before or simultaneously with the leaves (B) and are borne singly, or in umbels of two to six, occasionally more (C). The fruit is a fleshy drupe with a relatively large, single stone (D).[25]

Prunus avium, as shown here, is a deciduous tree growing up to ~29m bearing long-stalked flowers grouped into clusters. Fruits are long-stalked, small versions of the cultivated cherry, dark red to blackish in colour.

Habitat

In a natural setting, *Prunus avium* can be found at the margins of deciduous woodland. It is native to western Europe but has become naturalised throughout north America, New Zealand and Australia.

Associated insect benefits

Prunus species are an important early forage source for insects, with flowers emerging early in the year. Primary forage is pollen as nectar yields from this species are relatively poor (26-50 kg/Ha at best).[5,38] When nectar is available, it is reported to form a light and delicate honey which granulates rapidly with a fine grain structure.[5] Pollen from *Prunus* species has a crude protein content of 25-38%;[2,39] those at the higher end of this range are considered among the most nutritionally valuable.

Various cultivars exist as ornamentals, often with multi-layered flowers which are of more limited use for insects as the nectaries and stamens are less accessible.

Bee forage potential[1]		Notable foraging bee species[1]
Nectar/pollen	N/P	Early solitary species; red mason bee
Honey bees	✿✿✿	*Osmia rufa*, tawny mining bee, *Andrena fulva*
Short-tongued bumble bees	✿✿	Ashy mining bee, *Andrena cineraria*
Long-tongued bumble bees	✿✿	Buff-tailed bumble bee, *Bombus terrestris*
Solitary bees	✿✿✿	

Pollen presentation

Pollen loads of *Prunus avium* are brown,[11] and pollen grains ~30-40 μm, oval-triangular in cross section with three colporate apertures. The surface structure is striated.[40] This species has a pollen coefficient of 25 and is normally represented in honey.[3]

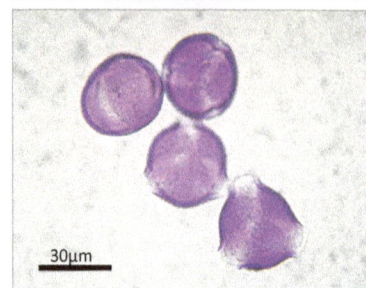

30μm

Rosaceae—Cherry Laurel

Family	Rosaceae
Species	*Prunus laurocerasus*
Common name	Laurel, cherry laurel
Flowering time	April-June

Species notes

Cherry laurel, *Prunus laurocerasus,* is a robust evergreen shrub commonly planted in parks and gardens. A native species to south-eastern Europe, it is now widely naturalised in UK woodlands. Trees reach ~15 metres.

Leaves are dark green, hairless, shiny and leathery, reaching up to 20-30 cm in length on a mature specimen. Sometimes bearing a few small teeth along the margins depending upon variety. *Prunus laurocerasus rotundifolia* (A) has broad, mid green blunt-ended leaves, while *Prunus laurocerasus Otto Luyken* (B), a more compact and slower growing variety, (B) has a dark green, highly glossy and sharp ended leaf.

Flowers are borne in erect racemes from early April onwards. Each flower is ~7-9 mm across with five petals and numerous long stamens (C). Fertilised flowers mature into shiny blue-black fruits 10-12 mm across. Cyanide compounds in this species give the leaves a distinct almond aroma when crushed.[25]

Habitat

Cherry laurel is widely naturalised and tolerant of most conditions other than extreme cold.

Associated insect benefits

Cherry laurel is attractive to bees and other pollinators; the flowers produce prolific nectar which is a valuable spring forage source. However it is not associated with a significant honey crop.[9]

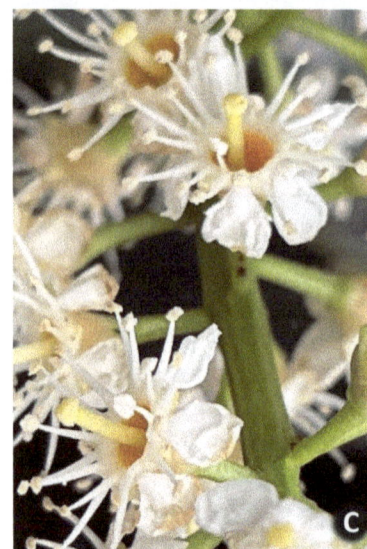

The plant has conspicuous extra-floral nectaries to the underside of the leaves, close to the midrib a short distance from the petiole (D). These provide nectar throughout the year, are worked by many insects, and are particularly attractive to honey bees and wasps.[41]

Bee forage potential[1]	
Nectar/pollen	N/p
Honey bees	✤✤
Short-tongued bumble bees	✤✤
Long-tongued bumble bees	✤
Solitary bees	✤✤

Notable foraging bee species[1]
Many *Bombus* species
Early solitary bees e.g. *Andrena fulva*
Honey bee, *Apis mellifera*

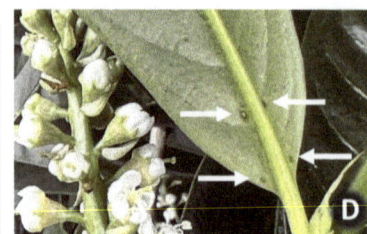

Pollen presentation

Pollen loads of *Prunus laurocerasus* are grey-green.[11] The pollen grains are ~35-40 μm and rounded-triangular in cross section with three colporate apertures and a striate surface texture.[42] The exine appears of medium thickness under light microscopy.

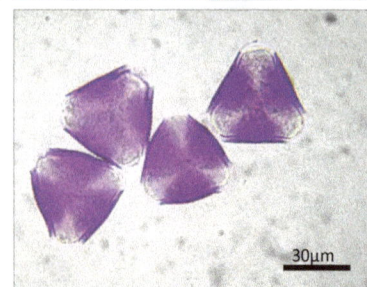

Rosaceae—Cherry Laurel

Prunus laurocerasus
Otto Luyken

Prunus laurocerasus
rotundifolia

Rosaceae—Apple

Family	Rosaceae
Species	*Malus pumila*
Common name	Apple
Flowering time	April-May

Species notes

Apple is a largely cultivated species, although not uncommon to be naturalised in hedgerows and waste ground where seeds are bird-sown or apple cored have been discarded. Trees grow to ~17 m and are ubiquitous throughout the UK with the exception of the far north of Scotland. Leaves are alternate, with finely serrated margins, grow to ~10 cm, are dull (not shiny) and may be downy on the underside only.[25]

Clusters of pale pink blossom appear in April-May (A,B); petals are 13-30 mm in length. Each cluster may contain four-six flowers.[9]

Cultivation has brought about over 7500 cultivars of apple with different attributes bred specifically for different tastes and purposes (C).[43]

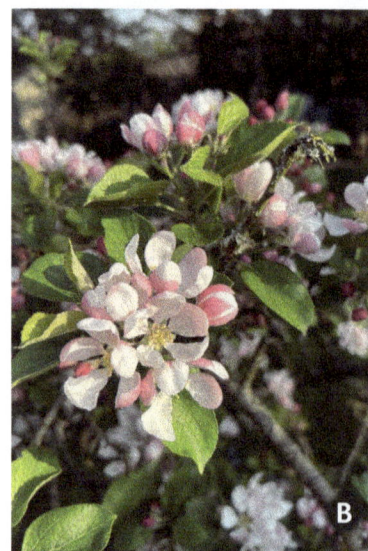

Habitat

Woodland scrub and hedgerows over fertile, humus-rich soil is preferred.[9]

Associated insect benefits

The open flowers of apple provide access and therefore forage for a wide range of bee species. Apples are self-incompatible, requiring cross pollination for fruit to develop. Spring flowering coupled with the scale of resources in orchard settings make apple a valuable forage source for both nectar and pollen. Although there is no significant honey yield, when honey is obtained, it is light amber in colour with good flavour and aroma.[1] Crude protein content of *Malus* spp. Is ~24%, indicating that the pollen is adequate to meet the protein requirements of a honey bee colony, but not exceptional.

Different cultivars may flower at slightly different times, offering a succession of forage for pollinators over many weeks. Fallen fruit also offers a valuable sugar source for wasps and hornets.

Bee forage potential[1]		Notable foraging bee species[1]
Nectar/pollen	N/P	Honey bee, *Apis mellifera*
Honey bees	✿✿✿	Many *Bombus* species
Short-tongued bumble bees	✿✿	Solitary bees including red mason bee
Long-tongued bumble bees	✿✿	*Osmia rufa,* tawny mining bee *Andrena*
Solitary bees	✿✿✿	*fulva* and ashy mining bee *A. cineraria*

Pollen presentation

Pollen loads are yellow-green.[3,11] The pollen grains are ~35-40 μm and rounded-triangular in cross section with three colporate apertures and a striate surface texture.[44] The exine appears of medium thickness under light microscopy.

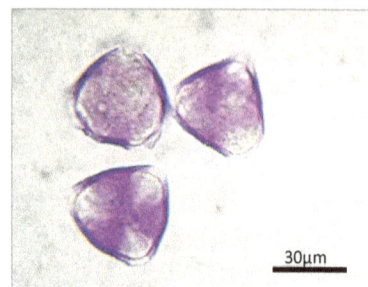

30μm

Rosaceae—Blackberry

Family	Rosaceae
Species	*Rubus fructiosa*
Common name	Blackberry, bramble
Flowering time	May-October

Species notes

Bramble is a ubiquitous thorny shrub producing stems that may reach 4m in length and can root from tips which arch to the ground, thereby producing substantial and impenetrable thickets. Stems are biannual or perennial, producing flowers in the second year.[25] Leaves typically comprise 5 leaflets (sometimes 3 or 7) on a thorny petiole.

White or pale pink flowers are produced from May through to October, an initial flush of flowers being followed by a lower level but consistent blossoming late into the summer. Each flower has 5 petals and a multitude of stamens (B,C).

Fruits are a composite of many single-seeded segments form in the late summer, ripening from green through red to glossy black (D).

Bramble is a complex of at least 334 apomictic microspecies i.e. those in which the seed develops directly from the ovule without the need for fertilisation. Resulting plants are therefore clones of the parent. Minor variations that occur because of mutations are passed on unchanged. Because bramble seeds are spread far and wide by birds and mammals, this can lead to populations of microspecies— genetically identical plants differing only in very minor characteristics. Some microspecies exist over a very wide area (e.g. spreading great distances via migrating birds), and are difficult to differentiate.[25,45]

Habitat

Found throughout the UK. Woodland, hedgerows, commons and scrub. Bramble is deep rooted, insensitive drought and tolerant of most soil conditions.[25]

Associated insect benefits

Bramble is a key forage species for many pollinating insects, including bees, butterflies, hoverflies and wasps.[36] It is an important honey crop for honey bees, producing a light and well-flavoured honey that is slow to granulate. Crane likens bramble honey to that from clover.[5] However the crude protein content of bramble pollen is only 14.8-20% which alone would not meet the protein requirements of a honey bee colony.[2] The open flowers are easily accessed by short-tongued species and a wide range of solitary bees and hover flies. Dried stems offer nesting habitat for small solitary bees including *Hylaeus* spp. and the blue carpenter bee (*Ceratina cyanea*) in south-east England.[1]

Bee forage potential[1]		Notable foraging bee species[1]
Nectar/pollen	N/P	Many *Bombus* species e.g. buff-tailed bumble bee
Honey bees	🐝🐝🐝	*B. terrestris*, white-tailed bumble bee *B. leucorum*,
Short-tongued bumble bees	🐝🐝🐝	tree bumble bee *B. hypnorum*.
Long-tongued bumble bees	🐝🐝🐝	Honey bee, *Apis mellifera*
Solitary bees	🐝🐝🐝	Solitary bees *Andrena, Colletes, Lassioglossum* spp.

Pollen presentation

Pollen loads are a dull grey-green.[10,11] Pollen grains are ~30 μm and generally oval-triangular in cross section, although can be highly variable. There are 3 colporate apertures and a gently striate surface texture.[46] The exine is of medium thickness, sometimes showing spaced rods.[10] This species has a pollen coefficient of 50 and is normally represented in honey.[3]

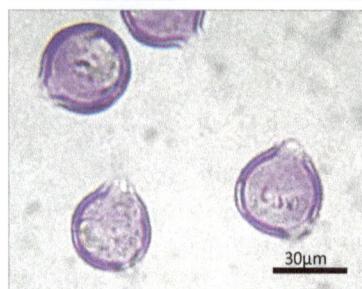

30μm

Rosaceae—Raspberry

Family	Rosaceae
Species	*Rubus idaeus*
Common name	Raspberry
Flowering time	May-August

Species notes

Raspberry, *Rubus idaeus*, is a perennial plant which bears biennial stems from a perennial root stock. Stems ('canes') are erect, reaching 1.5 m, and have a whitish bloom. Leaves have 5-7 oval leaflets and are white and woolly underneath (A). Green-white flowers are produced on second-year stems in small clusters, often drooping. Flowers have 5 delicate white petals and many stamens (B,C). Fruit, usually red but rarely yellow, are slightly downy and pull easily from the receptacle when ripe (D).[25]

Habitat

Raspberry is common throughout the UK, mostly abundant in the North and West. Preferred habitat is woods, heathland scrub, verges and rocky places. Can appear spontaneously from bird-sown seeds.

Associated insect benefits

Raspberry flowers are highly attractive to a wide range of pollinators, including solitary bees, short– and long-tongued bumble bees and honey bees. The quantity of fruit produced is directly related to quality of pollination; bumble bees are the more effective pollinators of this species, which may be related to the 'buzz pollination' method.[1,9]

The downward-hanging flowers produce prolific nectar, which is protected from rainfall by the orientation of the flower, making raspberry a valuable forage source following rainfall. Raspberry provides a reasonable nectar yield and a minor honey crop from honey bees, yielding 50-100 kg/Ha. The rapidly granulating honey is light with a fine, mild flavour.[1,5]

Although bees forage for pollen from raspberry, the pollen is poor quality, with a crude protein content of 19.0-21.3%, which in isolation, would be less than that required to meet the pollen requirements of a honey bee colony.[2]

Bee forage potential[1]		Notable foraging bee species[1]
Nectar/pollen	N/P	Buff-tailed bumble bee, *Bombus terrestris*
Honey bees	✿✿✿	Early bumble bee, *Bombus pratorum*
Short-tongued bumble bees	✿✿✿	Tree bumble bee, *Bombus hypnorum*
Long-tongued bumble bees	✿✿	Red mason bee, *Osmia rufa*
Solitary bees	✿	

Pollen presentation

Pollen loads are a light grey-green.[11] Pollen grains are ~30 μm and generally oval-triangular in cross section with 3 colporate apertures, a thin exine and a gently striate surface texture.[47]

This species has a pollen coefficient of 50 and is normally represented in honey.[3]

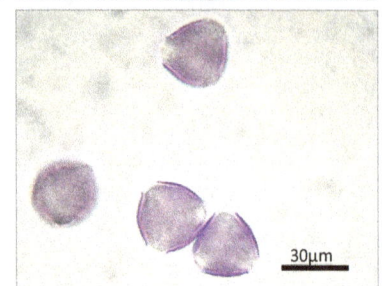

Rosaceae—Cotoneaster

Family	Rosaceae
Species	*Cotoneaster* spp.
Common name	Cotoneaster
Flowering time	May-June

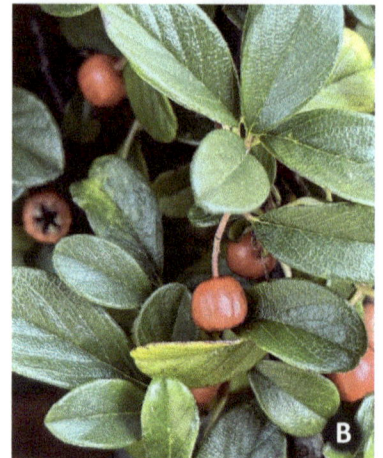

Species notes

Cotoneaster is a genus of flowering plants native to temperate Asia, Europe and north Africa, now widely naturalised in the UK, particularly in scrubby terrain. Some are invasive and may be a threat to native vegetation.[25] Depending upon the definition used, there may be anything from 70-300 individual species. Like bramble, cotoneaster has many apomictic microspecies which are treated as species or varieties depending upon the source. The majority are shrubs from 0.5-5 m in height, varying from prostrate plants to erect tree-like shrubs. Both evergreen and deciduous species occur. Leaves are generally alternate, 0.5-15 cm long, ovate to lanceolate depending upon species. Those pictured here are (A) *Cotoneaster horizontalis* and (B) *Cotoneaster dammeri*—both low-growing evergreen shrubs with a spreading habit; and (C) *Cotoneaster salicifolius* (willow-leaved cotoneaster)—a tree-like specimen reaching several metres.[48]

Flowers, either solitary or borne in corymbs of up to 100, are variable from white through to deep pink, with five petals, 10-20 stamens and up to five styles.[9]

Habitat

Cotoneaster is generally hardy and once established will thrive in most soil conditions, including at high altitude.

Associated insect benefits

Cotoneaster is a rich nectar source and the flowers are highly attractive to many different pollinators. The flat, open flowers are accessible to both long– and short-tongued bees. Many cotoneaster species flower during June, when other forage species may be absent, thus providing a welcome source of nutrients when there may be a dearth in some areas.[49]

Cotoneaster species are also used as larval food plants by some Lepidoptera species including grey dagger (*Acronicta grisea*), mottled umber (*Erannis defoliaria*), short-cloaked moth (*Nola cucullatella*), winter moth (*Operophtera brumata*), and hawthorn moth (*Scythropia crataegella*).

Bee forage potential[1]		Notable foraging bee species[1]
Nectar/pollen	N/p	Early bumble bee *Bombus pratorum*
Honey bees	✿ ✿ ✿	Red-tailed bumble bee, *Bombus lapidarius*
Short-tongued bumble bees	✿ ✿ ✿	Buff-tailed bumble bee, *Bombus terrestris*
Long-tongued bumble bees	✿ ✿	White-tailed bumble bee, *Bombus leucorum*
Solitary bees	✿ ✿	Honey bee, *Apis mellifera* and *Andrena* spp.

Pollen presentation

Pollen loads are yellow-green.[9] Pollen grains are generally ~25-30 µm, triangular with three colporate apertures, though this is variable by species—sometimes four.[50] The surface texture is gently striate, but appearing largely smooth under light microscopy.

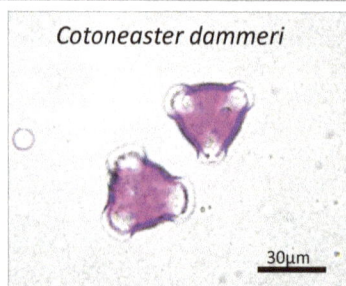

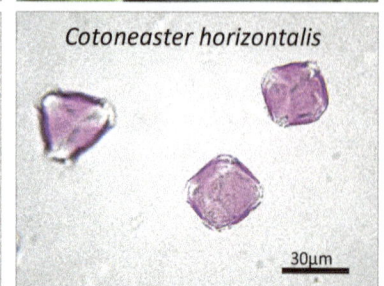

Cotoneaster dammeri

30µm

Cotoneaster horizontalis

30µm

Cotoneaster salicifolius

Cotoneaster francetii

Salicaceae—Willow

Salicaceae includes 56 genera and 1220 species, all of which are trees or shrubs with simple alternate leaves. Typically these species are fast-growing. The family includes willows (genus *Salix*), and also poplar, aspen and cottonwoods (genus *Populous*). Most species in temperate regions are deciduous and all are dioecious with flowers often arranged in catkins or spikes. They are often among the first plants to colonise disturbed, areas especially in wetlands.

Family	Salicaceae
Species	*Salix* spp. *Various species*
Common name	Goat willow, crack willow, white willow
Flowering time	March onward depending on species

Species notes

A number of Salicaceae species are important for pollinators. Probably foremost among these is the goat willow, *Salix caprea,* pictured (A-D). Goat willow forms a small tree, up to 10 m. Leaves are broad relative to most willow species (i.e. *Salix alba* [white willow] and *Salix fragilis* [crack willow] which display a more typical long, thin leaf), with slightly wavy margins, small teeth and very prominent veins.[25] Leaves are hairless above and slightly downy underneath. Male catkins are stout and covered with silky grey fur, erupting with copious pollen when ripe. Female catkins are slimmer and green. Hybridisation is common among willow species, and this can make species identification challenging.

Habitat

Most willow species occur predominantly in wet/damp environments—riverbanks, lake shores etc. *S. caprea* is more tolerant of drier ground than some willow species. The species occurs throughout northern and central Europe.

Associated insect benefits

Goat willow foliage is eaten by the caterpillars of a number of moths, including the sallow kitten, (*Furcula furcula*) sallow clearwing (*Synanthedon flaviventris*), dusky clearwing (*Paranthrene tabaniformis*) and lunar hornet clearwing (*Sesia bembeciformis*). It is also the main food plant for the purple emperor butterfly.[51,52] Catkins provide an important source of early pollen and nectar for bees, and are attractive to honey bees, solitary and bumble bees. While both male and female catkins have nectaries,[51] the male plant has been shown to be significantly more attractive to bees; this may be due to greater scent emitted[9] or the greater visual stimulus of the bright yellow pollen compared with the slightly duller green female catkin.[53]

There is no significant nectar presentation from willow, although Crane suggests a potential yield of 101-200 kg/Ha from the family generically and where honey can be obtained, it is light amber with a mild flavour and fine aroma.[5]

Predominant benefit is as a pollen source, particularly *S. caprea* which has a crude protein percentage of 36.8% - among the highest of any common forage species[2] which coupled with its early presentation makes this one of the most valuable forage species for bees.

Bee forage potential[1]		Notable foraging bee species[1]
Nectar/pollen	N/P	Many *Bombus* species, esp. early queens
Honey bees	🌼🌼🌼	Honey bee, *Apis mellifera*
Short-tongued bumble bees	🌼🌼🌼	Mining bees, *Colletes cunicularius* (rare
Long-tongued bumble bees	🌼🌼🌼	species), *Andrena clarkella*
Solitary bees	🌼🌼🌼	*Anthrophora* and *Osmia* spp.

Pollen presentation

Pollen loads from *S. caprea* are yellow.[11] The pollen grains are ~20 μm and generally round to oval in cross section with 3 colporate apertures, a thin exine and a reticulate/netted surface texture.[10,54] *S. caprea* has a pollen coefficient of 36.8 and would be normally represented in honey. *S. alba* and *S. fragilis* have coefficients of 17 and 14.8-15.1 respectively and would be under-represented in honey.[2]

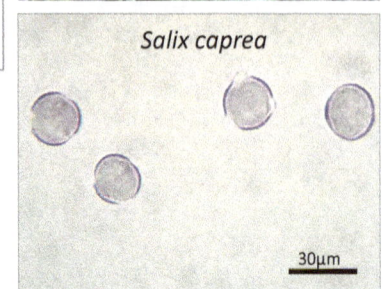

Salix caprea

30μm

Salix fragilis

Salicaceae—Willow

Salix alba

Salix caprea

Lythraceae—Purple Loosestrife

Lythraceae is a family of flowering plants including 32 genera, with about 620 species of herbs, shrubs and trees. The species is generally characterised by simple leaves with an opposite (sometimes whorled or alternate) arrangement and pinnate venation. Flowers are bisexual with 4, 6 or 8 sepals and an equal number of petals. Typically twice as many stamens as petals.[9]

Family	Lythraceae
Species	*Lythrum salicaria*
Common name	Purple loosestrife
Flowering time	June-September

Species notes

Purple loosestrife (*Lythrum salicaria*) is a robust herbaceous perennial which can grow up to 1-1.5 m in height. Numerous erect stems originate from a single woody root mass. The stems are reddish-purple with distinctive square cross-section (A). Leaves are lanceolate, 3–10 cm long and 5–15 mm across, downy and sessile, and arranged opposite or in whorls of three. Flowers are purple, 10–20 mm in diameter, with six petals (occasionally five) and 12 stamens, and are clustered tightly in the axils of bracts or leaves; It is native to Europe and Asia.[8,9]

There are three different flower types, with the stamens and style of different lengths; short, (in which the organs are completely concealed within the floral tube) medium (in which the organs are visible at the mouth of the floral tube) and long (in which the organs protrude beyond the floral tube. This species possesses trimorphic incompatibility—each flower type can only be pollinated by one of the other types, not the same type, thus ensuring cross-pollination between different plants.[8,55]

Habitat

Purple loosestrife is generally found in moist ground; fenland, wet meadows, river/canal banks and lakesides (C). It tends not to thrive in acid soils.[25]

Associated insect benefits

Purple loosestrife is highly beneficial to insects, producing both pollen and nectar in large quantities, though there is no significant honey crop in the UK. When available the honey is said to be dark and strongly flavoured.[1]

Pollination is generally by long-tongued insects including solitary bees, bumble bees, honey bees and butterflies. In addition to bees, a number of other insects use purple loosestrife as a food source, including beetles (e.g. the black-margined loosestrife beetle *Galerucella calmariensis* and loosestrife root weevil *Hylobius transversovittatus*).

Bee forage potential[1]		Notable foraging bee species[1]
Nectar/pollen	N/P	Many *Bombus* species
Honey bees	✿ ✿	Esp. Common carder bee, *Bombus*
Short-tongued bumble bees	✿	*pascuorum*
Long-tongued bumble bees	✿ ✿	Honey bee, *Apis mellifera*
Solitary bees	✿	

Pollen presentation

Pollen loads are yellow-green to deep green.[11] Pollen grains are ~20-25 µm in diameter however this varies with flower type—those with long style have small pollen grains (~15 µm); those with short style long filament have larger pollen grains (~25 µm). Pollen grains are round but triangular cross section with smooth slightly striated surface structure. There are three colporate apertures.[56]

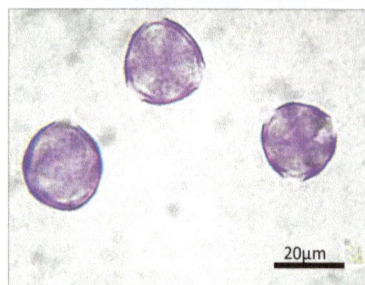

20µm

Onagraceae—Willowherb

Onagraceae or willowherb family (sometimes evening primrose family) is a family of ~650 species of herbs, shrubs and trees across 17 genera. The family is widespread, occurring in both tropical and temperate climates. The family is characterised by flowers which generally have four sepals and petals. Leaves are commonly opposite or whorled. Many members of this family are distinguished by a characteristic four-parted stigma.[9]

Family	Onagraceae
Species	*Chamaenerion angustifolium*
Common name	Rosebay willowherb, fireweed (US)
Flowering time	June-September

Species notes

Rosebay willowherb, *Chamaenerion angustifolium* is a perennial herbaceous flowering plant in the willowherb family Onagraceae. It is also known by the synonyms of *Chamerion angustifolium* and *Epilobium angustifolium*. Its reddish stems with spirally arranged alternate lanceolate and stemless leaves can reach 1.5 m in height.

The inflorescence is a symmetrical terminal raceme (A) which blooms sequentially from bottom to top, with each fertilised flower forming a long, thin seed capsule that can contain up to 400 seeds, such that each plant may produce as many as 80,000 seeds (C). Flowers are deep cerise, ~20 mm across, with petals of unequal size (B). Rosebay willowherb is protandrous (the male flower parts mature before the female) thus favouring cross-pollination. Once established, the plant can also spread by means of an extensive root system. As such, rosebay willowherb can form large stands of dense growth.[25]

Habitat

Rosebay willowherb is ubiquitous throughout the British Isles. It is a pioneer species, and will quickly colonise open spaces and disturbed ground, particularly after fire.

Associated insect benefits

The flowers are visited by a wide variety of insects including a number of Lepidoptera for whom it is their primary larval host-plant; examples including the elephant hawk moth (*Deilephila elpenor*),[57] bedstraw hawk moth (*Hyles gallii*), and the white-lined sphinx moth (*Hyles lineata*).[58]

Its long and sequential flowering period is highly beneficial to pollinating insects. Rosebay willowherb produces ample nectar and pollen and is attractive to both short and long-tongued bumble bees and honey bees. This species has a honey potential of 5-6, yielding potentially over 500 kg of honey/Ha under optimal conditions. Honey from rosebay willowherb is very pale with a fine flavour, and will granulate with a fine grain structure.[5]

Although plentiful, the pollen is of poor quality, with a mean protein content of 16.2%.[2]

Bee forage potential[1]		Notable foraging bee species[1]
Nectar/pollen	N/P	Buff-tailed bumble bee, *Bombus terrestris;*
Honey bees	✿ ✿ ✿	White-tailed bumble bee, *Bombus lucorum;*
Short-tongued bumble bees	✿ ✿ ✿	Common carder bee, *Bombus pascuorum*
Long-tongued bumble bees	✿ ✿ ✿	Honey bee, *Apis mellifera*
Solitary bees	✿	Solitary species including. *Lasioglossum* spp.

Pollen presentation

Pollen loads are characteristically blue-grey.[11] Pollen grains large, at ~70 μm in diameter. There are three pores with thickened edges, giving the round grains a triangular appearance.[59] The exine is of medium thickness; grain contents can appear granular. The pollen grains also posses fine viscin threads that are visible under light microscopy. These are thought to aid aggregation of the pollen grains and adhesion to pollinating species.[60] It has a pollen coefficient of 0.3 and is greatly under-represented in honey.[3]

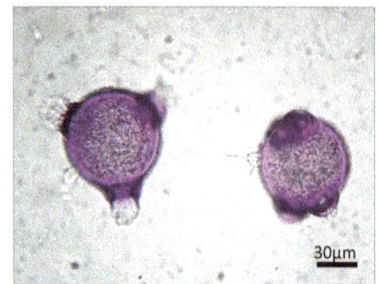

30μm

Onagraceae—Willowherb

Sapindaceae—Horse Chestnut

Sapindaceae is a family of flowering plants sometimes known as the maple or soapberry family. It contains 145 genera and 1,925 species occurring mainly in the tropical or subtropical regions; a small number thrive in temperate climates. Most are woody trees, large shrubs and woody climbers with insect pollinated flowers. The family includes a number of key species for pollinators, including horse chestnut, sycamore and maple.

Family	Sapindaceae
Species	*Aesculus hippocastanum*
Common name	Horse chestnut, conker tree
Flowering time	May-June

Species notes

Horse chestnut, *Aesculus hippocastanum,* is a large and robust flowering deciduous tree growing to about 39 m (A). It is widely naturalised across much of the northern hemisphere as far north as the Arctic circle. It has large leaves, divided palmately into 5-7 leaflets, each up to 30 cm long in a mature specimen. [25,61]

Flowers are borne in conspicuous upright panicles 15-30 cm high. Each bears 20-50 individual blooms, with flowers having five sepals and 4-5 white petals with a coloured blotch at the base (B). This is coloured yellow in unpollinated flowers turning to pink once the blossom is pollinated (C). This is noteworthy as bees do not see reds, so the absence of visible nectar guide signals the lack of reward at a pollinated blossom. Three types of flower exist: male only (the majority), female only and hermaphrodite (which tend to occur toward the base of the panicle). [1]

Usually one to five fruits develop on each panicle. The shell is a green, spiky capsule containing one (rarely two or three) nut-like seeds (conkers) (D).

Habitat

Horse chestnut is tolerant of many environments and will thrive in most temperate areas provided summers are not too hot or ground too waterlogged.

Associated insect benefits

Horse chestnut produces copious pollen and nectar; the size and lengthy flowering period of the tree makes for a valuable forage source for bees. [1] Flowers are largely insect pollinated.

The sticky leaf buds of this species are a source of propolis for honey bees.

Caterpillars of the triangle moth (*Trigonodes hyppasia*) feed on the leaves of horse chestnut. [62]

Bee forage potential[1]		Notable foraging bee species[1]
Nectar/pollen	N/P	Many *Bombus* species
Honey bees	✿ ✿ ✿	Honey bee, *Apis mellifera*
Short-tongued bumble bees	✿ ✿	
Long-tongued bumble bees	✿ ✿	
Solitary bees	✿	

Pollen presentation

Pollen loads are terracotta red. [11] Pollen grains are ~20 μm in diameter, elongated round in shape, with three colporate apertures and a smooth, thin exine. [63] Granules or projections can be seen to ornament the apertures. [10,63]

Aesculus hippocastanum has a pollen coefficient of 26.7; the pollen is normally represented in honey. [2]

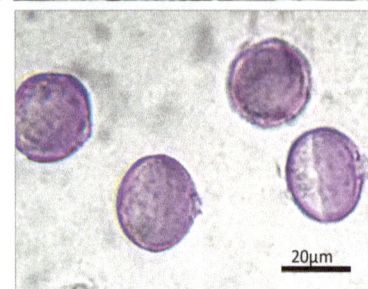

20μm

Sapindaceae—Horse Chestnut

Sapindaceae—Maple

Family	Sapindaceae
Species	*Acer* spp. e.g. *Acer platanoides*
Common name	Maple, Japanese Maple, Norway Maple
Flowering time	April

Species notes

Acer is a significant genus of trees and shrubs, commonly known as maples. There are approximately 123 species, most native to Asia. Most are substantial deciduous trees growing up to 45 m. Maples have a characteristic opposite leaf arrangement, the leaves being palmate, veined and lobed with 3-9 veins (occasionally 13) leading to each lobe (A).

Flowers are borne in racemes, corymbs or umbels, and have four or five sepals, four or five petals about 1–6 mm long (absent in some species), four to ten stamens about 6–10 mm long, and two pistils or a pistil with two styles again, depending on species (B,C).[64] Distinctive fruits or 'samaras' occur in pairs with an attached fibrous 'wing' which carries on the wind (D)

Many ornamental cultivars are available, some with highly decorative leaves, autumn colour or flower colour (C).

Habitat

Maples grow in a range of habitats and at varying altitudes but prefer deep, moist, fertile soils. They are a major constituent of many temperate forests.[65]

Associated insect benefits

Maple provides a reasonable supply of nectar and pollen, especially for early bees as well as a variety of flies and other insects. The accessible flowers are worked by short and long-tongued bumble bees, solitary bees and honey bees. There is no notable honey crop.

Maple pollen can have a high crude protein content—21.4-39.4% depending on variety, the red maple, *Acer rubrum*, being at the upper end of this range. It can therefore be highly beneficial to insects.[2]

Those worked for nectar include Norway maple, *Acer platanoides* (shown here in B and pressed sample), Italian maple, *Acer opalus*, and sugar maple, *Acer saccharum*.[1]

Bee forage potential[1]		Notable foraging bee species[1]
Nectar/pollen	N/P	Mainly beneficial to early-foraging
Honey bees	✿ ✿	species:
Short-tongued bumble bees	✿	Early bumble bee, *Bombus pratorum*
Long-tongued bumble bees	✿	Tawny mining bee, *Andrena fulva*
Solitary bees	✿	

Pollen presentation

Pollen loads are yellowish or light though dark green to brownish, depending upon species.[11] Pollen grains are ~30 μm in diameter, oval/triangular in shape, with three sunken colpate apertures and a smooth, medium exine. Small striated detail may be seen at high magnification.[10,66]

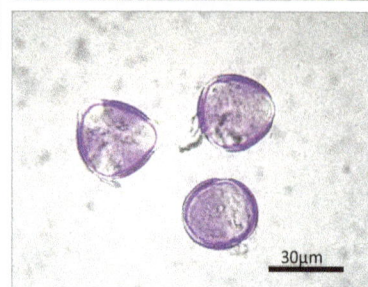

30μm

Sapindaceae—Sycamore

Family	Sapindaceae
Species	*Acer pseudoplatanus*
Common name	Sycamore
Flowering time	April-June

Species notes

Sycamore is a large, deciduous broad-leaved tree reaching ~35 m.[25] Sycamore was introduced to the UK in the late 15th century, and although not native to the UK, it is widely naturalised thanks to its ability to self-seed in almost any habitat. It is considered an invasive species in some territories.[67]

The tree has large five-pointed leaves with coarsely toothed lobes and red leaf stalks. Flowering occurs when the leaves are newly expanded.[67] Green-yellow flowers are borne in groups of three in long pendent racemes or panicles of up to 16cm, each bearing up to 100 flowers (A,B). Each flower is monoecious, 4-6 mm pale green or yellowish, with short petals. Males have whitish stamens and bright yellow anthers. Usually, the centre flower is female, and develops into the seed (C), while the two lateral flowers are male, and produce pollen.[1]

Habitat

Sycamore will grow in virtually any soil type provided there is adequate drainage, although a dry, free soil is preferred over one that is stiff or moist.[1]

Associated insect benefits

Sycamore is highly beneficial to insects. Although flowers are small, they are plentiful and yield ample nectar and pollen, attracting many kinds of pollinators.

Sycamore can yield an amber-coloured honey, sometimes greenish, the flavour of which has been described as both 'unremarkable' and 'rank to some'. It is slow to granulate with a fine grain structure.[1,5] Sycamore trees are frequently heavy with sycamore aphids, *Drepanosiphum platanoidis,* which secrete honeydew, also collected by honey bees. This can darken the honey and endow a characteristic honeydew flavour.

Bee forage potential[1]		Notable foraging bee species[1]
Nectar/pollen	N/P	Many *Bombus* species
Honey bees	✿ ✿ ✿	Honey bee, *Apis mellifera*
Short-tongued bumble bees	✿ ✿	Some solitary bees including *Andrena*
Long-tongued bumble bees	✿ ✿	*bucephala* (rare, southern England and
Solitary bees	✿ ✿	Wales)

Pollen presentation

Sycamore pollen loads are greenish grey.[5] Pollen grains are ~30-40 μm in diameter, oval/triangular in shape, with three colpate apertures and a smooth, medium exine. Small striated detail may be seen at high magnification.[10,68]

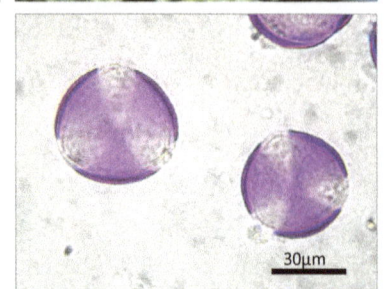

30μm

Malvaceae—Lime

Malvaceae is a diverse family of flowering plants containing some 244 genera and 4225 known species. The family includes trees such as lime and cacao as well as ornamentals such as hollyhock, mallow and hibiscus. Although quite varied, many have characteristically broad, slightly hairy alternate leaves and five-petalled flowers associated with conspicuous bracts. Most are entomophilous, avoiding self pollination via protandry.

Family	Malvaceae
Species	*Tilia* spp. e.g. *Tilia x europaea; Tilia platyphyllos*
Common name	Lime, Linden (Eur), Basswood (US)
Flowering time	June-July

Species notes

Lime, *Tilia* spp. is a genus of about 30 deciduous trees and bushes native throughout the temperate northern hemisphere. Most commonly in the UK is the common lime, *Tilia x europaea*; the large-leaved lime *Tilia platyphyllos* is also widely naturalised. Both are substantial trees that may reach 40 m in height.[25]

Leaves are large, heart shaped and with finely toothed margins. The green-coloured flowers are found in pendulous clusters of 4-10 blossoms attached to a greenish bract (various stages shown, A-D).[25]

Habitat

Lime will grow and thrive in virtually any soil type.[1] It is frequently found as a planted specimen in parkland and urban settings.

Associated insect benefits

Lime is highly beneficial to insects, producing ample pollen and copious quantities of nectar at temperatures of ~20°C and above. Large trees can give heavy nectar crops estimated at over 500 kg/Ha, resulting in a light, sometimes slightly greenish honey with a characteristic flavour.[5] However the nectar supply is said to be fickle, dependent up on suitably warm and humid weather.[9] Pollen is also plentiful, however it is of relatively low quality, with a protein content of 18.2%.[5]

Lime is highly susceptible to aphid infestation, making it highly attractive to ladybirds, the adults and larvae of which predate aphids. As such, lime trees are also a source of honeydew which is foraged by a multitude of insects.

Bee forage potential[1]		Notable foraging bee species[1]
Nectar/pollen	N/P	Many *Bombus* species including
Honey bees	⊛ ⊛ ⊛	Buff-tailed bumble bee, *Bombus terrestris*
Short-tongued bumble bees	⊛ ⊛ ⊛	Red-tailed bumble bee, *Bombus lapidarius*
Long-tongued bumble bees	⊛ ⊛	Honey bee, *Apis mellifera*
Solitary bees	⊛	

Pollen presentation

Pollen loads from Lime are yellow-brown.[11] Pollen grains are medium sized, ~40 μm in diameter, with a convex triangular cross section and three colporate apertures with a distinctive aperture appearance under light microscopy. The exine is of medium thickness, with a slightly reticulated surface structure.[10,69]

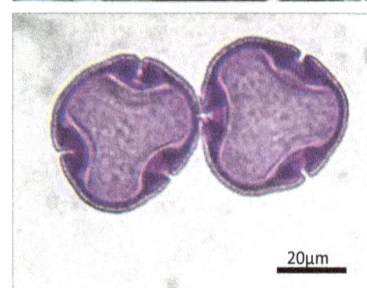

20μm

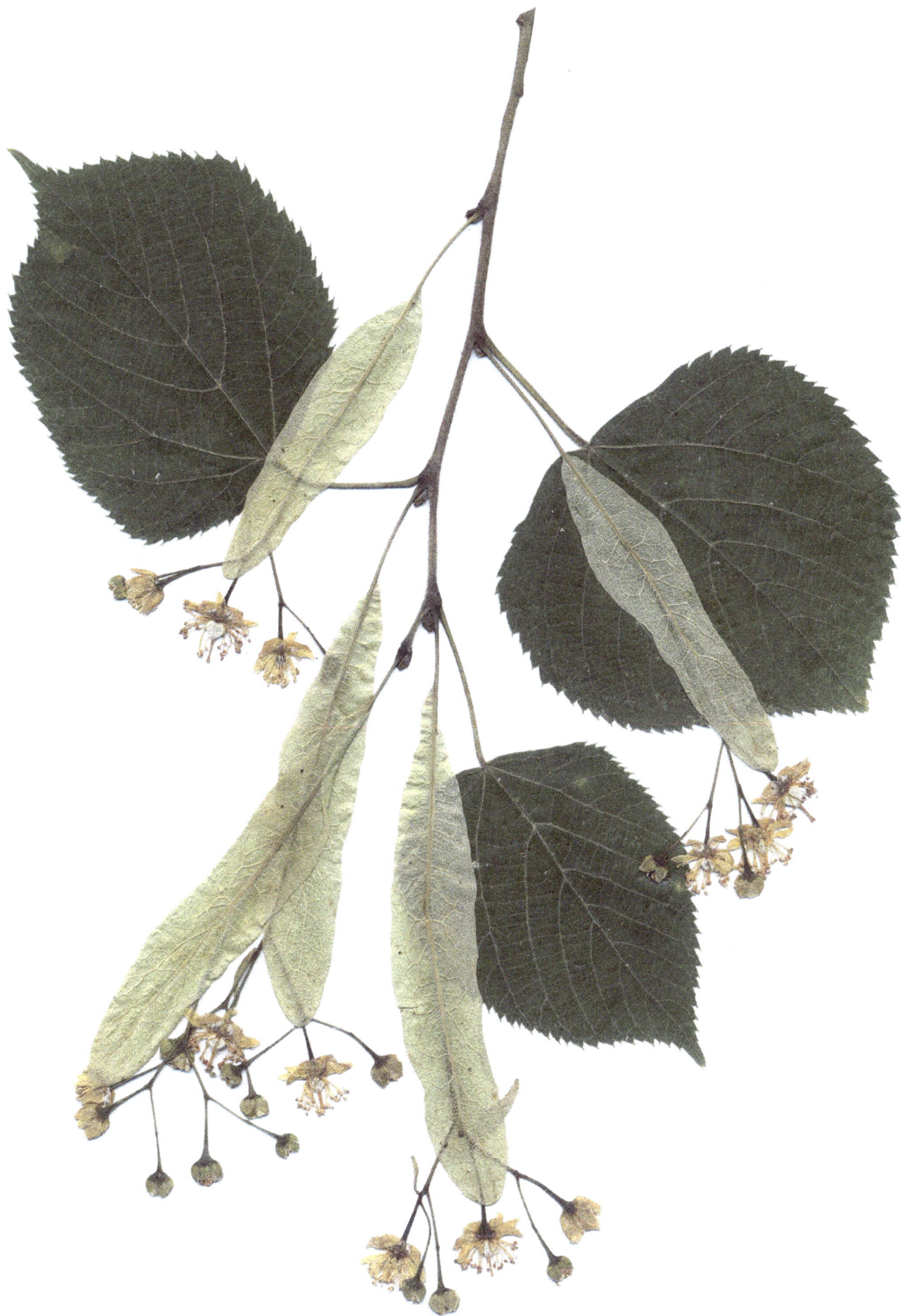

Brassicaceae—Oil Seed Rape

Brassicaceae is a family of flowering plants also known as mustards or crucifers (on account of their four-petalled flowers). Of 4060 species (372 genera) most are annual, biennial or perennial herbaceous plants. Leaves are generally simple, appearing alternately or in rosettes. The family includes many economically important vegetables and crop species, including cultivated cabbage, kale, broccoli, turnip and radish.

Family	Brassicaceae
Species	*Brassica napus*
Common name	Oil seed rape
Flowering time	April-May (winter sown)

Species notes

Oil seed rape (*Brassica napus*) is a widely cultivated slightly waxy and somewhat glaucous annual to biennial plant reaching some 1m in height. Lower leaves can be sparsely hairy with a large terminal lobe; upper leaves are unlobed and clasp the stem (A). Flowers are bright yellow, and about 17mm across, consisting of four petals alternating with four sepals, and held in a dome of blossoms with buds slightly over-topping the open flowers (B).[25] *Brassica napus* is predominantly cultivated in its winter-sown form in Europe and Asia as it requires vernalisation to initiate flowering.

Habitat

This species can be cultivated on a wide variety of soils unless waterlogged, although there is a preference for neutral to slightly alkaline soil conditions.[70]

Associated insect benefits

Although oil seed rape is largely anemophilous, insect pollination can increase seed set, and it is a major honey-yielding forage species in the UK, producing up to 500 kg/Ha,[5] although temperatures need to exceed ~16°C for the plant to secrete nectar. Nectaries are at the base of the flower, which may be difficult to reach for honey bees, however they overcome this by accessing the nectaries between the base of the petals—a technique termed 'base working'.[9] Honey is pale and delicately flavoured, its high glucose content leading to rapid small-grained granulation. In addition to honey bees, *Eristalis* hoverflies and bumble bees, particularly *Bombus lapidarius* are key pollinators for this species.[71] The pollen is nutritionally valuable, with a crude protein content of 22.8-31.9%.[2]

Bee forage potential[1]		Notable foraging bee species[1]
Nectar/pollen	N/P	Honey bee, *Apis mellifera*
Honey bees	✿ ✿ ✿	Buff-tailed bumble bee, *Bombus terrestris*
Short-tongued bumble bees	✿ ✿ ✿	Red-tailed bumble bee, *Bombus lapidarius*
Long-tongued bumble bees	✿ ✿ ✿	Common carder bee, *Bombus pascuorum*
Solitary bees	✿ ✿	Solitary bees: *Andrena* and *Lasioglossum* spp

Pollen presentation

Pollen loads are yellow. Pollen grains are ~30 μm in diameter, round but can appear to have a triangular cross section when viewed from the polar orientation; the exine has a slightly netted surface structure and is of medium thickness with spaced rods.[8] There are three colpate apertures.[72] Oil seed rape has a pollen coefficient of 150 and is over-represented in honey.[3]

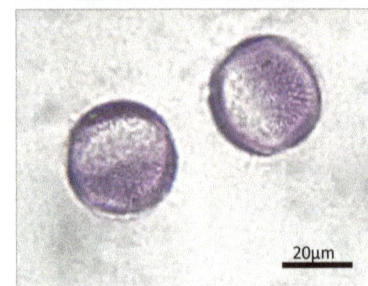

20μm

Brassicaceae—Oil Seed Rape

Plumbaginaceae—Sea Lavender

Plumbaginaceae is a diverse family of flowering plants, mostly perennial herbaceous species and a few shrubs found across many climatic regions. They are particularly associated with saline environments, including salt-rich steppes, brackish marshes and sea coasts. The family includes 21 genera and ~725 species. The majority are *Limonium* or sea lavender genus, accounting for ~600 species.[73]

Family	Plumbaginaceae
Species	*Limonium vulgare*
Common name	Sea lavender
Flowering time	July-early October

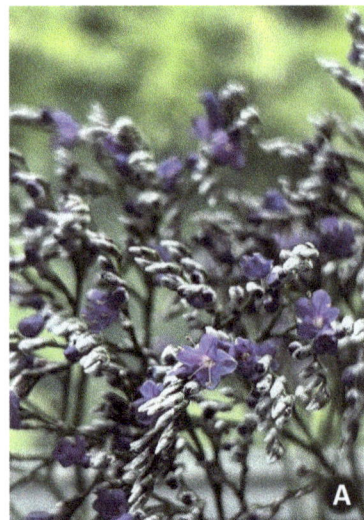

Species notes

Common sea lavender *Limonium vulgare*, is a locally abundant perennial plant reaching ~40 cm in height; more frequently smaller. The fleshy leaves form a loose and partially upright basal rosette (B). Sprays of small purplish flowers are produced on a panicle held on stiff, waxy stems. Flowers are composed of five papery petals with white papery bracts to the outside of the flower. Each flower has five prominent stamens (C). *Limonium vulgare* will frequently hybridise with Lax-flowered sea lavender (*Limonium humile*) making unambiguous differentiation difficult except by microscopic analysis of plant and pollen.[25]

Habitat

Sea lavender is a locally abundant perennial plant of coastal and inter-tidal salt marshes, and occasionally shingle and sea walls. Its preference for saline environments dictates its coastal distribution under natural circumstances.

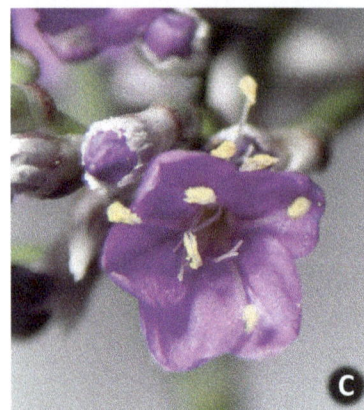

Associated insect benefits

Sea lavender is often prevalent in remote places where other sources of forage may be limited. It is a useful source of nectar late in the season when many other forage species are over.[1] In areas of prevalence, this species may yield a honey crop of light coloured honey of good quality although quantities are relatively small—Crane suggests only up to 50 kg/Ha.[5] Pollen is also foraged by honey bees, however its quality is poor, with a percentage crude protein content of 16%.[2]

Bee forage potential[1]		Notable foraging bee species[1]
Nectar/pollen	N/p	Some Bombus species and locally prevalent pollinators
Honey bees	✿ ✿	Mostly honey bee, *Apis mellifera*
Short-tongued bumble bees	✿	
Long-tongued bumble bees	✿	
Solitary bees	✿	

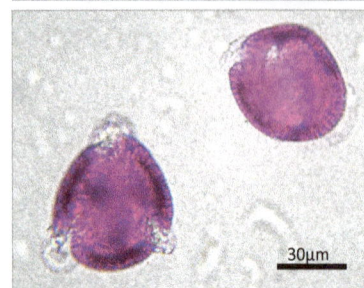

Pollen presentation

Pollen loads are pale yellow. Pollen grains are ~50 μm in diameter, round but triangular in cross section with a visibly thick exine and deeply netted surface structure. There are three colpate apertures at which the bulging intine is visible.[74]

Balsaminaceae—Himalayan Balsam

Balsaminaceae is a family of annual or perennial plants falling into two genera—*Impatiens* (>1000 species) and *Hydrocera* (1 species). Plants of this family are found throughout temperate and tropical regions, primarily in Asia and Africa, but also North America and Europe.[75] Most *Impatiens* species are herbaceous plants with succulent stems and protandrous flowers showing a wide variety of flower forms.

Family	Balsaminaceae
Species	*Impatiens glandulifera*
Common name	Himalayan balsam, Indian balsam
Flowering time	July-October

Species notes

Himalayan balsam, *Impatiens glandulifera*, is a large annual plant reaching 2 m in height in substantial stands. Not native to the UK, it was introduced in 1839 as an ornamental, and spread rapidly, primarily along water courses in the first instance. The plant has a soft greenish red angular stem and lanceolate and finely toothed leaves up to 15 cm in length held in whorls of three (A,B). Extra-floral nectaries are present below the leaf stems. Flowers range from very pale through dark pink (C), and have a characteristic hooded shape.[25] After flowering, seed pods form which are 2-3 cm in length (D)—when ripe, these rupture spontaneously or when touched, propelling seeds explosively up to 7 m.[25]

Habitat

Himalayan balsam is mostly found along riverbanks, and in wet woodland and marshes. Its native range is at altitudes of 200-2500 m above sea level, however in the UK, is it found in lowland settings up to 320 m.

Associated insect benefits

Although an invasive non-native, Himalayan balsam is an excellent source of forage for most bees and also wasps from late summer through to the first frosts. Foraging insects enter the tube of the flower to access nectar from a narrow, curved spur at the back of the flower. In doing so they brush against the overhead stamen becoming coated in a characteristic pattern of white pollen. In 2014, *I. glandulifera* was rated highest in the top ten UK plants in terms of sugar production per flower per day.[76] When foraged by honey bees, this species can produce a heavy honey yield of medium well flavoured honey.

Bee forage potential[1]		Notable foraging bee species[1]
Nectar/pollen	N/P	Most bee species, but frequently:
Honey bees	✿ ✿ ✿	Honey bee, *Apis mellifera*
Short-tongued bumble bees	✿ ✿ ✿	Buff tailed bumble bee, *Bombus terrestris*
Long-tongued bumble bees	✿ ✿ ✿	Common carder bee, *Bombus pascuorum*
Solitary bees	✿	

Pollen presentation

Pollen loads are grey-white.[10] Pollen grains are characteristically oblate, ~20x35 μm, with four colpate apertures. The thin exine has a very finely reticulate surface structure which appears smooth under light microscopy.[77]

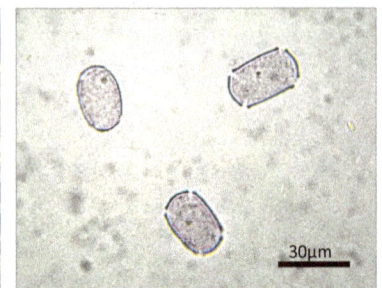

30μm

Balsaminaceae—Himalayan Balsam

53

Ericaceae—Bell Heather

Ericaceae is a family of flowering plants including the heaths and heathers, and also rhododendron and azalea and a number of species cultivated for their fruits; cranberry, blueberry and huckleberry. Most commonly found in acidic and otherwise infertile growing conditions, the family includes 124 genera and 4250 species of generally evergreen herbs, shrubs and trees.

Family	Ericaceae
Species	*Erica cinerea*
Common name	Bell heather
Flowering time	June-October

Species notes

Bell heather, *Erica cinerea*, is probably more accurately termed bell heath, as it is more closely related to heath (*Erica*) than to ling heather (*Calluna*). It is a low-growing bushy evergreen with dark green stiff foliage and purplish coloured, urn-shaped flowers which are notably larger and brighter in colour than Ling heather (A). Leaves are in whorls of three, very short stalked and 4-7 mm in length, with their white undersides mainly obscured by the curled longitudinal edges. Flower clusters are borne at the top of the stems (B), each blossom consisting of a 4-6 mm corolla which splits into four curved teeth at the tip. Each has eight stamens which are held inside the corolla.[25]

Habitat

A native to the west of Europe, Bell heather is found in a variety of habitats, particularly heathland, open woodland, and some coastal areas. It thrives best in acidic, well-drained soils (C).

Associated insect benefits

Bell heather is an excellent source of nectar and pollen for insects; the corolla is the right length for foraging by honey bees and other short-tongued species.[1] Honey bees may gather a surplus which results in a brownish honey with a 'port-wine– colour and characteristic flavour.[5] Ruby tiger moths (*Phreagmatobia lineata*) and silver-studded blue butterflies (*Plebejus argus*) also frequent this species for nectar. It was rated in the top five for most nectar production (nectar per unit cover per year) survey of UK plants conducted in 2014.[76]

Bee forage potential[1]		Notable foraging bee species[1]
Nectar/pollen	N/P	Honey bee, *Apis mellifera*
Honey bees	✿ ✿ ✿	White-tailed bumble bee, *Bombus lucorum*
Short-tongued bumble bees	✿ ✿ ✿	Garden bumble bee, *Bombus hortorum*
Long-tongued bumble bees	✿ ✿ ✿	Heath bumble bee, *Bombus jonellus*
Solitary bees	✿ ✿ ✿	Heathland mining bee, *Colletes succinctus*

Pollen presentation

Pollen loads are grey-white.[11] Pollen grains are characteristic tetrads but larger than many of the other *Erica* species at ~50 μm in diameter, with three colporate apertures.[78] The exine is smooth, and of medium thickness.[10] *Erica cinerea* has a pollen coefficient of 10, and is under-represented in honey.[3]

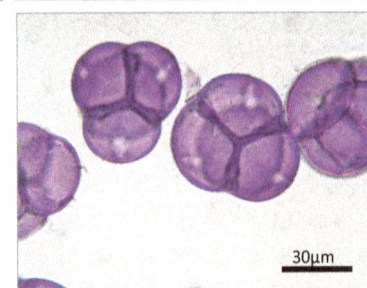

30μm

Ericaceae—Heaths

Family	Ericaceae
Species	*Erica carnea, Erica tetralix*
Common name	Winter heath, cross-leaved heath
Flowering time	December-March; June-October

Species notes

The genus *Erica* includes some 800 low growing bushy evergreen shrubs. Most are indigenous to South Africa, but a number are found in more northern regions. Notable and most common in the family in the UK are winter heath (***Erica carnea***) (A-C), and cross-leaved heath, (***Erica tetralix***) (D). *Erica carnea* is typified by needle-like leaves 4-8 mm long, borne in whorls of four. Flowers are a slender bell shape 4-6 mm long, usually light through dark pink, but occasionally white. Each is formed from five fused petals forming a bell-shaped corolla. The stamens and pistils extend from the mouth of the corolla .The plant flowers profusely through the winter.[79]

Erica tetralix is a summer flowering variant. This has leaves of ~2-5 mm in cross-shaped whorls of 4 and more urn-shaped flowers that droop from the ends of the shoots. Flower tubes are ~7-8 mm in length and lack the protruding stamens of *E. carnea*.[25]

Habitat

Erica carnea is native to mountainous areas of central , eastern and southern Europe, although in the UK it is valued and widespread as a cultivar because of its winter flowering. Unlike most varieties of *Erica*, this species will tolerate mildly alkaline as well as acidic soils. *Erica tetralix* is a native of more western areas, from southern Portugal to central Norway, where it is found in acid soils of blanket bog and uplands.[25]

Associated insect benefits

Both of these *Erica* species are excellent forage species for pollinators. The short flowers of *E. carnea* in particular are accessible to long and short-tongued bees.[1] Honey bees can gather a surplus from this species, yielding an amber coloured honey that may be sharply flavoured.[5]

With slightly longer flowers, *E. tetralix* is more valuable to bumble bees as well as some solitary and mining bees.[1]

Bee forage potential[1]		Notable foraging bee species[1]
Nectar/pollen	N/P	Honey bee, *Apis mellifera*
Honey bees	⊛ ⊛ ⊛	White tailed bumble bee, *Bombus lucorum*
Short-tongued bumble bees	⊛ ⊛ ⊛	Garden bumble bee, *Bombus hortorum*
Long-tongued bumble bees	⊛ ⊛ ⊛	Heath bumble bee, *Bombus jonellus*
Solitary bees	⊛ ⊛ ⊛	Heathland mining bee, *Colletes succinctus*

Pollen presentation (*Erica carnea*)

Pollen loads are grey-brown.[11] Pollen grains are characteristic tetrads and ~30 µm in diameter, with three colporate apertures.[80] The exine is smooth, and of medium thickness.[10] *Erica carnea* has a pollen coefficient of 10, and is under-represented in honey.[3]

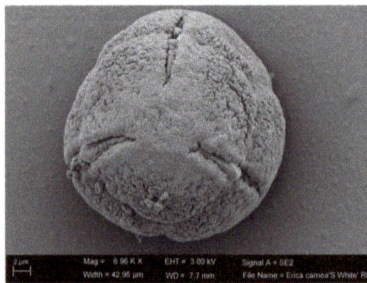

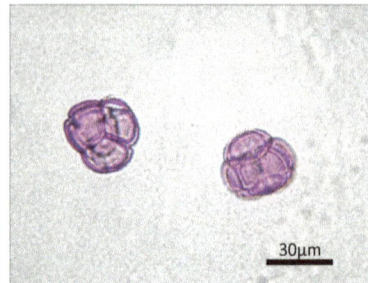

30µm

Ericaceae—Ling Heather

Family	Ericaceae
Species	*Calluna vulgaris*
Common name	Ling heather
Flowering time	August-September

Species notes

Ling heather, *Calluna vulgaris*, the only species in the genus *Calluna,* is a low-growing bushy evergreen with tiny leaves (~2-3 mm) that are packed closely, and grow oppositely along the stems (B). It reaches ~60 cm. Sprays of tiny purple-pink flowers, rarely white, appear in late summer. The base of each flower is enclosed by tiny purple-green bracts. The calyx is split into four purplish sepals, all of which are longer than the corolla, which is 4-5mm long and split into four petals; there are eight stamens.[25]

Habitat

Ling heather native to Europe, and is found as far north as the Faroe Islands and Iceland. It has been introduced more widely to areas of suitable habitat including N America, Australia, New Zealand and the Falkland Islands. It is found preferentially in poor, acid soils, mostly either sandy or peaty ground including moor and heathland (D).[81]

Associated insect benefits

Ling heather is considered a valuable forage for honey bees and a variety of bumble bees and solitary bees. As a honey-producing species, it is considered a premium forage, yielding a characteristically strong flavoured honey that varies in colour from light through dark to reddish brown, and is thixotropic.[5] The nectar of this species contains a megastigmane, callunene, which, at naturally occurring concentrations, has been shown to inhibit the trypanosome *Crithidia bombi,* a common parasite of bumble bees.[82] Some Lepidoptera species, including the small heath butterfly (*Coenonympha pamphilus*) and the Scotch argus butterfly (*Erebia aethiops*), may also visit heather plants for nectar. The UK Pollinator Monitoring Scheme (PoMS) survey reported in 2024 and others found that primary visitors to heathers were Diptera—flies and hoverflies, as well as honey bees and bumble bees.[36,83,84]

Bee forage potential[1]		Notable foraging bee species[1]
Nectar/pollen	N/P	Honey bee, *Apis mellifera*
Honey bees	✿✿✿	White-tailed bumble bee, *Bombus lucorum*
Short-tongued bumble bees	✿✿✿	Heathland mining bee, *Colletes succinctus*
Long-tongued bumble bees	✿✿✿	*Andrena fuscipes; Nomada rufipes* (lays
Solitary bees	✿✿✿	eggs in nest of *A. fuscipes*)

Pollen presentation

Pollen loads are grey-brown.[11] Pollen grains are medium sized, arranged in tetrads characteristic of the family—these are ~32 µm in diameter. The exine is smooth, and of medium thickness.[10] There are three colporate apertures.[85] *Calluna vulgaris* has a pollen coefficient of 12 and is under-represented in honey.[3]

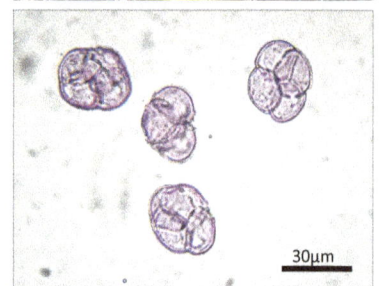

30µm

Ericaceae—Rhododendron

Family	Ericaceae
Species	*Rhododendrum ponticum*
Common name	Rhododendron
Flowering time	May-June

Species notes

The **Rhododendron** genus in the largest in the *Ericaceae* family, including some 1024 species, most of which are native to eastern Asia. **Rhododendron ponticum** was introduced to the UK as a cultivar in 1763 and has since become widely naturalised. It is a large, robust woody evergreen shrub reaching ~5 m in height, and which can form dense thickets that exclude most other species. Leathery, hairless oval to oblong leaves reach ~2 cm in mature specimens. Showy purple flowers are borne in impressive and conspicuous clusters of bell-shaped blooms, each ~4-6 cm across, at the end of the branches. Flowers typically have 5 petals, fused at the base, and 10 prominent stamens.[25] The fruit is a dry capsule 1.5 to 2.5 cm (0.59 to 0.98 in) long, containing numerous tiny, wind-dispersed seeds. Each flower head can produce between three and seven thousand seeds, so that a large bush can produce several million seeds per year.[86]

Habitat

Rhododendrum ponticum is a woodland species which thrives on acid soils, where it will regenerate freely from seed.[87,88]

Associated insect benefits

Although rhododendron flowers are open and accessible, and pollen is readily accessible, it is rarely foraged by honey bees, and it is primarily visited by bumble bees. Flowers produce a large quantity of sugar-rich nectar (~30%),[88] making it attractive to insects. The nectar and pollen of rhododendron contains acetylandromedol, a grayanotoxin, which can be toxic to honey bees; bumble bees are unaffected.[1, 89]

A number of Lepidoptera species are known to forage rhododendron, including swallowtail (*Papilio*) spp. Similarly the shrub is frequented by a number of common hoverfly species and beetles. In a study reported in 2006, Stout and colleagues observed various generalist pollinators visiting this species, predominantly *Bombus* spp., also other Hymenoptera and hoverflies (*Syrphidae* spp).[88]

Bee forage potential[1]		Notable foraging bee species[1]
Nectar/pollen	N/P	Early bumble bee, *Bombus pratorum*
Honey bees	–	White-tailed bumble bee, *Bombus lucorum*
Short-tongued bumble bees	✿✿✿	Bilberry bumble bee, *Bombus monticola*
Long-tongued bumble bees	✿✿✿	Red-tailed bumble bee, *Bombus lapidairus*
Solitary bees	–	

Pollen presentation

Pollen grains are characteristic tetrads and ~60 μm in diameter, with three colporate apertures. The exine is smooth, and appears of medium thickness under light microscopy.

The anthers of *Rhododendrum ponticum* do not dehisce along their length, as is usual. Rather, an aperture opens at the distal end of the anther and pollen disperses through the hole. Gathering sufficient for pollen analysis can be challenging!

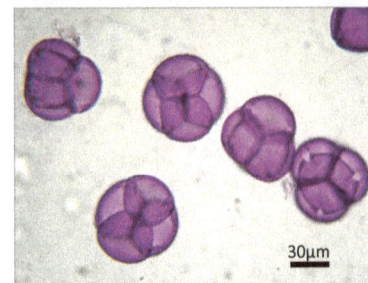

30μm

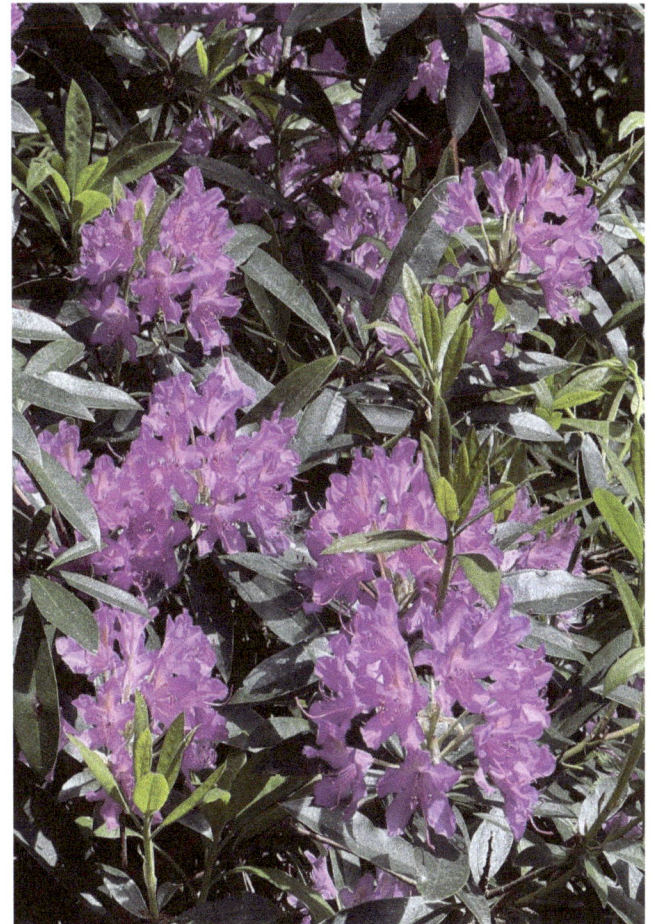

Boraginaceae—Borage

Boraginaceae is a wide ranging family of 2000 species of shrubs, trees and herbaceous plants covering ~150+ genera. Most members of the family have characteristically hairy leaves and monoecious flowers (a small number are dioecious). The family includes species that are among the most valuable for pollinators and particularly honey bees, including borage, phacelia and vipers bugloss.

Family	Boraginaceae
Species	*Borago officinalis*
Common name	Borage, starflower
Flowering time	June-September

Species notes

Borage, *Borago officinalis*, is a roughly hairy annual (A,B) with characteristic blue flowers (genetically recessive white variants may exist as cultivars).[90] The basal rosette of prominently wrinkled leaves are stalked and oval, reaching ~15 cm. Stem leaves are smaller and stalkless, sometimes with a slightly wavy edge. The plant reaches ~60 cm.[25]

The pendulous flowers are ~20 mm across, star-shaped, with five triangular-pointed petals (C). The five-lobed calyx is usually extremely hairy. Five dark coloured prominent anthers are central to the corolla.

Habitat

Although native to the Mediterranean, borage will thrive in UK lowland areas, and is tolerant of most soil types. It will self-seed readily.[25]

Associated insect benefits

Borage is a highly beneficial plant for pollinators, and is worked freely by honey bees and bumble bees species. Some, particularly *B. leucorum* and *B. terrestris* buzz the anthers, becoming covered in the pollen which they are then able to groom and collect.[1]

Nectar is secreted at the base of the ovary, and collected as bees insert the proboscis between the stamens.[1] Honey bees will readily collect a surplus from borage, particularly where it is grown commercially for its oil content. The resulting honey is light and delicately flavoured.[5]

Bee forage potential[1]		Notable foraging bee species[1]
Nectar/pollen	N/P	Honey bee, *Apis mellifera*
Honey bees	✿✿✿	White tailed bumble bee, *Bombus lucorum*
Short-tongued bumble bees	✿✿✿	Buff-tailed bumble bee, *Bombus terrestris*
Long-tongued bumble bees	✿✿✿	Common carder bee, *Bombus pascuorum*
Solitary bees	✿✿	Early bumble bee, *Bombus pratorum*

Pollen presentation

Pollen loads are grey-white.[11] Pollen grains are ~30 μm in diameter, round in cross section with smooth exine of medium thickness.[10] There are frequently more than six colporate apertures.[91]

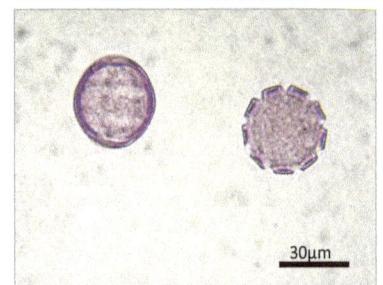

30μm

Boraginaceae—Forget-me-not

Family	Boraginaceae
Species	*Myosotis* spp.
Common name	Forget-me-not
Flowering time	April-October depending on species

Species notes

Myosotis is a genus of annual, herbaceous flowering plants including over 500 recorded species, the majority of which occur in western Eurasia and New Zealand. Many are now common in temperate areas following their introduction as cultivars. Flowers are typically <1 cm across, flat-faced and coloured blue or pink with white or yellow centres (A,B). Foliage is alternate (C). Most species typically flower in spring, and many have lengthy flowering periods.[25]

Forget-me-not as we know it relates to a sub-group of the genus, including **Myosotis arvensis,** (field forget-me-not; most abundant in the UK, shown here), *Myosotis sylvatica* (wood forget-me-not) and others, all of which are low-growing annuals or perennials with small flowers (~3-5 mm); details vary by species.

Habitat

Myositis are most common as cultivars or garden escapees, although will thrive in many areas and soil types, particularly moist river and stream banks, and woodland edges.[25]

Associated insect benefits

Early flowering myosotis species are a valuable nectar and pollen source at a time of the year when other forage may be scarce. Many bee species forage for nectar and pollen. The mouth of the flower is very narrow, and as bees insert the proboscis to access nectar within, they dislodge a great deal of pollen from the anthers, which sit just below the mouth of the flower tube. The pollen grains are tiny, and may reach the honey crop of the bee in much greater quantities than those of other species, resulting in a disproportionate prevalence of *Myosotis* spp. pollen in honey.[1] The nectar guide, in the centre of the corolla is coloured yellow in fresh flowers, fading to white in those that have been pollinated, thus signalling to pollinators that there is no nectar available (C).

Myosotis spp. Are commonly worked by a broad range of pollinators including bees, butterflies and flies.[92] It does not yield a honey crop from honey bees.[1]

Bee forage potential[1]		Notable foraging bee species[1]
Nectar/pollen	N/P	Honey bee, *Apis mellifera*
Honey bees	✿ ✿	Mining bees, *Andrena* spp.
Short-tongued bumble bees	✿ ✿	Cuckoo bees, *Nomada* spp.
Long-tongued bumble bees	✿ ✿	
Solitary bees	✿ ✿	

Pollen presentation

Pollen loads from *Myositis* are yellow.[10] The pollen grains are very small (<5 µm) in an elongated dumbbell shape. The exine is thin with a smooth surface texture; there are three colporate apertures,[93] but the size of the grain makes these impossible to see under normal light microscopy. Forget-me-not pollen has a pollen coefficient of 5000 and is greatly over-represented in honey.[3]

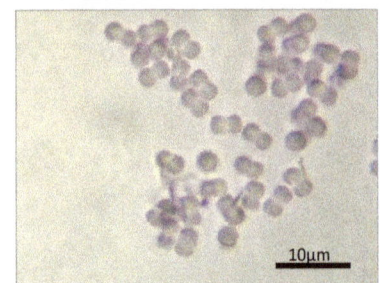

10µm

Boraginaceae—Viper's Bugloss

Family	Boraginaceae
Species	*Echium vulgare*
Common name	Viper's bugloss, blueweed (US)
Flowering time	June-September

Species notes

Viper's bugloss, *Echium vulgare*, is a coarsely hairy biennial plant native to Europe and central Asia. Reaching 80-100 cm in height, the vegetation starts initially as a basal rosette with simple oblanceolate leaves. These become progressively smaller toward the top of the flowering stems. Showy flowers, ~1-2 cm across, form in a helical configuration up the flower stalks. Flowers are pink when in bud, (A), maturing to blue as they open. The corolla of each flower is divided into five unequal lobes, and there are 5 projecting pink stamens of varying lengths.[25,94]

Habitat

Echium vulgare is common on rough and disturbed ground, particularly dry soils over sand, chalk and limestone. More recently its value to pollinators has been recognised and it is often found as a component of wild flower mix in naturalised planting which takes it outside of its natural habitat, which in the UK, is somewhat localised in the south.

Associated insect benefits

This species is highly valuable to pollinators. It produces nectar very freely and the open flower forms are accessible to both long and short-tongued bees as well as a host of other pollinators including the long hoverfly *Sphaerophoria scripta*.[1,95] The flowers are protandrous, and a study reported in 1990 found that the plants offer a higher nectar reward, and attract more pollinators during the male phase compared with the female phase.[95]

Nectar yields are not only high in quantity, but the quality of the nectar is good— Crane reports a sugar value of 0.09-1,64 mg sugar per flower per 24 h (average 1.3); among the highest of those noted.[5] The pollen from this species is also highly nutritious, with a crude protein content of 25.9-37.4%.[2] The plant may yield 500 kg/ha of nectar under optimum conditions. It can produce a pale to light golden honey with a delicate flavour.[5]

Bee forage potential[1]		Notable foraging bee species[1]
Nectar/pollen	N/P	Honey bee, *Apis mellifera*
Honey bees	⚘ ⚘ ⚘	Buff-tailed bumble bee, *Bombus terrestris*
Short-tongued bumble bees	⚘ ⚘ ⚘	Common carder bee, *Bombus pascuorum*
Long-tongued bumble bees	⚘ ⚘ ⚘	Great yellow bumble bee, *Bombus distinguendus*
Solitary bees	⚘ ⚘	Larger solitary species, *Anthrophora* spp

Pollen presentation

Pollen loads from viper's bugloss are dark blue-grey.[10,11] The pollen grains are very small—~15-20 μm— and prolate (slightly lengthened along the polar axis) with a thin section, smooth exine. There are three colporate apertures.[96]

This species has a pollen coefficient of 250 and is over-represented in honey.[3]

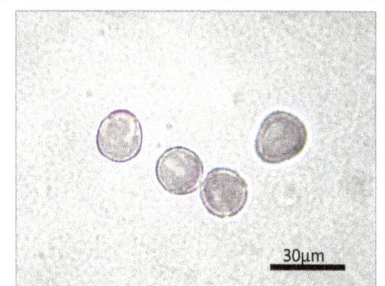

30μm

Boraginaceae—Phacelia

Family	Boraginaceae
Species	*Phacelia tanacetifolia*
Common name	Phacelia, blue tansy, purple tansy or fiddleneck
Flowering time	August-September

Species notes

Phacelia tanacetifolium, (commonly known as phacelia) is an upright annual that will generally reach ~70 cm. It is glandular, and slightly hairy. Leaves form a basal rosette with smaller leaves up the stem. Most are divided into smaller leaflets which are deeply cut and intricately toothed (A). The plant bears distinctive coiled spikes of mauve flowers with prominent projecting stamens and a forked style (B,C). Each blossom is 6-10 mm across, with the calyx cut almost to the base into 5 lobes forming a bell-shaped corolla.[25]

Habitat

Phacelia tanacetifolia is native to North America and was introduced to the UK as a cultivar in 1832. It was first recorded in the wild in 1885.[25] It is often sown as a green manure or in wildflower margins at field edges. Because of its value to pollinators, it also forms a substantial proportion of wild-flower seed mixes and hence can frequently be seen outside of its natural setting, appearing on roadside verges and set-aside land. It will thrive in virtually any soil type.[1]

Associated insect benefits

Phacelia is highly valuable to pollinators, and the flowers are particularly attractive to honey bees and short-tongued bumble bees.[1] Nectar is secreted freely and in large quantities from a disc at the base of the ovary;[1] Crane rates this species among the highest yielding, letting up to 500 kg/ha. When foraged by honey bees, it can result in an amber coloured glucose-rich honey that granulates quickly to a near-white set.[1,5]

The flowers are also a good source of pollen for hoverflies (Syrphidae).[97] Pollen from this species is nutritionally valuable, with a crude protein content of 28.1%.[2]

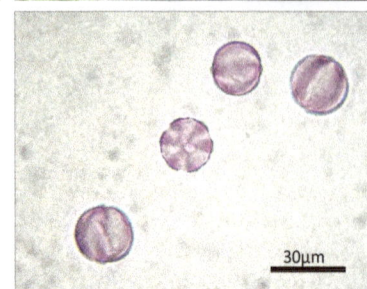

Bee forage potential[1]		Notable foraging bee species[1]
Nectar/pollen	N/P	Honey bee, *Apis mellifera*
Honey bees	❀ ❀ ❀	White-tailed bumble bee, *Bombus leucorum*
Short-tongued bumble bees	❀ ❀	Buff-tailed bumble bee, *Bombus terrestris*
Long-tongued bumble bees	❀	Solitary bees, *Halictus* and *Lassioglossum* spp.
Solitary bees	❀	

Pollen presentation

Pollen loads from this species are dark blue.[11] The pollen grains are ~25 μm, with three colporate apertures.[98] The light micrograph reveals grains which initially appear to have six apertures; however close inspection reveals three of these to be pseudocolpi between the actual apertures.[98]

30μm

Oleaceae—Privet

Oleaceae is a widely distributed family of flowering shrubs and trees comprising 28 genera and ~700 species. Most are woody trees plants, a few are lianas (uses others as a means of support, e.g. as seen in clematis). The family is characterised by opposite leaves. Evergreen species predominate in temperate and tropical regions; deciduous species are more common in colder climates.

Family	Oleaceae
Species	*Ligustrum vulgare, Ligustrum ovalifolium*
Common name	Privet
Flowering time	August-September; June-July

Species notes

Privet, is a flowering plant in the genus **Ligustrum**; the species **Ligustrum vulgare** is native to central and Europe, north Africa and Asia. It is a semi-evergreen, small shrub reaching ~3 m. Leaves are borne in opposite pairs characteristic of the family. Leaves are semi-shiny, 3-6 cm long by 1-1.5 cm across with a regular lanceolate shape.

Sprays of pungent creamy flowers with a waxy texture are produced in mid summer (A,B). Individual flowers are 4-5 mm across, with a ~3 mm corolla tube splitting into 4 small petals; there are two protruding stamens and the style (C).[25]

In urban settings, *Ligustrum vulgare* is less common than *Ligustrum ovalifolium,* the garden privet, (shown here). *L. ovalifolium* is more evergreen, with larger, more broadly oval leaves, and a slightly longer corolla tube which is longer than the petal lobes.[25]

Habitat

Privet, particularly *L. ovalifolium,* is ubiquitous in the UK, and is tolerant of most soil types in lowland areas.

Associated insect benefits

The short flower tubes of privet are accessible to many generalist pollinators, including short and long-tongued bees. Nectar secretion can be copious, and highly attractive to honey bees. If there is a surplus, it will produce a dark and strongly flavoured honey which is said to be bitter and distasteful.[1] Vine weevils (*Otiorhyncus sulcatus*) are commonly found on privet in the summer, with adults feeding on the leaves, and larvae on the roots.

Bee forage potential[1]		Notable foraging bee species[1]
Nectar/pollen	N/P	Honey bee, *Apis mellifera*
Honey bees	✿✿	White tailed bumble bee, *Bombus lucorum*
Short-tongued bumble bees	✿✿	Red-tailed bumble bee, *Bombus lapidarius*
Long-tongued bumble bees	✿✿	
Solitary bees	✿	

Pollen presentation

Pollen loads are yellow-green.[11] Pollen grains are ~30 μm in diameter, round in cross section with visibly netted surface. The exine is thick, with coarse rods visible. There are three colporate apertures.[99]

Privet pollen has a pollen coefficient of 25 and is normally represented in honey.[3]

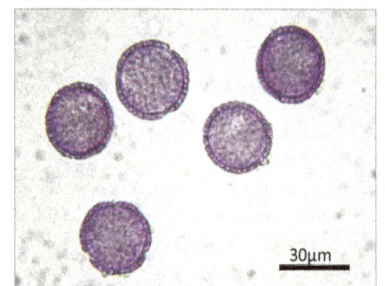

30μm

Lamiaceae—Thyme

Lamiaceae is a family of flowering plants containing many aromatic species that are used in food preparation, including thyme, mint, rosemary, oregano and marjoram. The family is widespread, and covers ~236 genera and up to 7500+ species. Typically, they have leaves arranged in opposite pairs, stems that have a square cross section and bilaterally symmetrical flowers with five fused petals.

Family	Lamiaceae
Species	*Thymus vulgaris, thymus polytrichus*
Common name	Thyme
Flowering time	May-August

Species notes

The genus **Thymus** contains ~350 species of aromatic perennials native to temperate climates. Native wild thyme in the UK, *thymus polytrichus* (B,C), is a dense, pungent, low-growing evergreen plant reaching ~10 cm. Small hairless or sparsely hairy oval leaves are arranged in opposite pairs along the reddish stems which will root at the nodes if they touch the ground. Flower stems are more square in cross section. Flowers are borne in dense terminal heads, comprising a multitude of small pink/purple flowers typical of the family configuration. The corolla is ~7 mm with four protruding stamens.[25]

Habitat

Thymus polytrichus is common throughout the UK, favouring chalk downland, upland grassland, cliff-tops and heath environments.[25]

Associated insect benefits

Both wild thyme and the cultivar *Thymus vulgaris* (A) are excellent forage plants and are well worked for nectar by a range of hoverflies, social and solitary bees.[1] Many of the *Thymus* family are highly rated for potential honey yield; Crane rates *Thymus vulgaris* as having the potential to yield over 500 kg per fully planted hectare.[5] The quantity of sugar produced per flower per 24 h is low (0.01-0.09 mg) though this is more a reflection of the small size of the flowers and a value normalised against flower size would be more useful.[1] Recent research has shown that both the nectar and more so, the pollen of *Thymus vulgaris* contain terpenes and acetates which are associated with antimicrobial activity, and have the potential to be protective against *Melissococcus plutonius* (European foul brood) if consumed in the hive. The concentration of leaf volatiles is higher still and it was postulated that this may be of benefit to leaf-cutter bees, specifically *Megachile rotundata* which is native to Eurasia and an introduced species in the US.[100]

Bee forage potential[1]		Notable foraging bee species[1]
Nectar/pollen	N/p	Honey bee, *Apis mellifera*
Honey bees	✿✿	Many solitary bees
Short-tongued bumble bees	✿✿	Some less common species e.g. blue
Long-tongued bumble bees	✿✿	mason bee, *Osmia caerulescens*
Solitary bees	✿✿	

Pollen presentation

Pollen loads are grey-white.[10] Pollen grains are ~30 μm in diameter, oval in cross section with six well-defined colpate apertures. The exine is of medium thickness, with a reticulate/netted surface texture.[10,101]

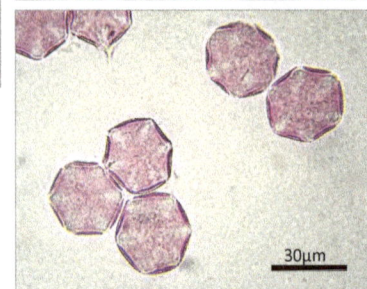

30μm

Lamiaceae—Mint

Family	Lamiaceae
Species	*Mentha* spp.
Common name	Mint (many varieties, naturalised and cultivar)
Flowering time	May-October depending on species

Species notes

Mints are a wide ranging group, prone to variation and hybridisation. Four are native to the UK—spear mint (**M. spicata**) (A, B), water mint (**M. aquatica**) (C), apple mint (*M. x villosa*) and pepper mint (*M. x piperata*). Many hybrids and cultivars exist. All are fragrant if bruised.

M. Spicata is the commonest garden species and is often found naturalised as an escapee. It has pointed stalkless, hairless leaves and characteristic purple stems. Terminal flower heads are in spikes, bearing a profusion of white-mauve flowers. It has a creeping habit via over-ground stolons and will readily establish from root fragments. *M. aquatica* is the most common wild species. It is a creeping and patch-forming perennial with opposite, sometimes slightly hairy, toothed leaves. Lilac flowers with projecting stamens are borne in dense rounded heads, often with smaller flower heads lower on the stem.[25]

Habitat

Most mint species can be found in many environments, but will grow best in moist soils.[25]

Associated insect benefits

Mint is highly beneficial for insects—the small flowers of these species are particularly attractive to short-tongued bees and honey bees, and also available to hoverflies, although long-tongued species will also visit, particularly those that are prevalent later in the summer. It is unfortunate that commercially grown crops are harvested before they flower. *M. spicata* and *M. x piperata* are sometimes cultivated on a field scale and, when foraged by honey bees, can yield a surplus of amber-coloured honey that has a slightly minty flavour, though this is said to diminish with time.[1] Crane notes the potentially high yield from these species at up to 500 kg/Ha, and also that honey from *M. aquatica* is unusual in that it contains vitamin C.[5]

Bee forage potential[1]		Notable foraging bee species[1]
Nectar/pollen	N	Honey bee, *Apis mellifera*
Honey bees	✿✿	Brown banded carder bee, *Bombus humilis*
Short-tongued bumble bees	✿	Common carder bee, *Bombus pascuorum*
Long-tongued bumble bees	✿	Tree bumble bee, *Bombus hypnorum*
Solitary bees	✿	

Pollen presentation

Pollen loads from *Mentha aquatica* (shown here) are yellow. The grains are ~30-35 μm in diameter with six colpate apertures. The exine is thin, with a reticulated surface.[102]

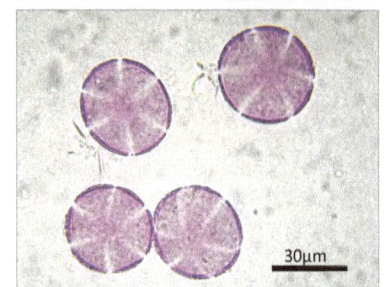

30μm

Lamiaceae—Lavender

Family	Lamiaceae
Species	*Lavandula angustifolia* and other species
Common name	Lavender
Flowering time	June-August

Species notes

Now a notable garden species in the UK, **Lavender, *Lavandula angustifolia*** is native to the Mediterranean area and was thought to be introduced during the Roman occupation as it was widely cultivated for its aromatic oils. The genus *Lavandula* includes ~30 species. It is a strongly aromatic shrub which can reach ~1 m in height. Evergreen leaves 2-6 cm long can be slightly glaucous. Purple flowers are sparsely arranged in spikes at the tip of bare, striated stalks which can reach ~30 cm.[103]

Flowers are of the form characteristic to the family, with a flower tube ~6mm in length. Small nutlet fruits are produced at maturity.[103]

Habitat

Lavender needs to be in full sun, and will thrive in light or chalky, well-drained soils and will tolerate dry conditions (but not damp) conditions which are similar to those of its native habitat.[1]

Associated insect benefits

Lavender is highly beneficial to insects. In its native range, it grows over large tracts of land and is an important honey plant, although this is not the case in the UK. Honey from *L. Spica* (broadleaved lavender) is golden to dark amber with a fine flavour that is notably reminiscent of the plant scent. It may have high water and sucrose content, and granulates with a fine grain structure.[5]

The flower tube of lavender is just short enough to permit access by short-tongued bees, and it is frequented by a large range of insects including a range of bumble bee species and solitary bees. The scarce green-eyed flower bee, *Anthrophora quadrimaculata,* has been noted visiting lavender and also cat mint in gardens within its range in southern UK.[1]

Although bees will gather pollen from lavender, it is of relatively poor nutritional quality, with a crude protein content of 19.4% (20-25 being considered the minimum to meet the nutritional needs of a honey bee colony).[2]

A relatively recent incomer to the UK, the rosemary beetle (*Chrysolina americana*), a handsome opalescent leaf beetle, is frequently found on lavender plants (B).

Bee forage potential[1]		Notable foraging bee species[1]
Nectar/pollen	N/P	Honey bee, *Apis mellifera*
Honey bees	❀ ❀ ❀	Red-tailed bumble bee, *Bombus lapidarius*
Short-tongued bumble bees	❀ ❀ ❀	Common carder bee, *Bombus pascuorum*
Long-tongued bumble bees	❀ ❀ ❀	Flower bees, *Anthrophora* spp.
Solitary bees	❀ ❀	

Pollen presentation

Pollen loads from lavender are yellow-orange.[104] The pollen grains are ~35-40 µm in diameter with six colpate apertures. The exine is thin-medium with a pitted surface texture.[105]

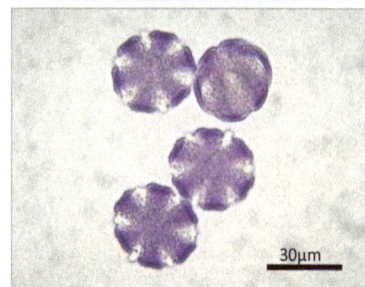

30µm

Lamiaceae—Rosemary

Family	Lamiaceae
Species	*Salvia rosmarinus*
Common name	Rosemary
Flowering time	April-June

Species notes

Rosemary, *Salvia rosmarinus,* is an aromatic evergreen shrub native to the Mediterranean region. However, it is cold-tolerant and is widely cultivated in the UK. There are number of different forms, depending on cultivar variety. The naturally occurring species is a robust upright woody shrub reaching ~1 m. Narrow opposite leaves ~1-2 cm in length are typically green above and white/woolly underneath (A). Small blueish tubular flowers are borne in axillary clusters. Each flower is small— 6-12 mm—comprising a two-lobed upper lip and three-lobed lower lip (B,C).[106]

Habitat

Rosemary thrives best in those conditions most reminiscent of its natural habitat—full sun, well-drained soil, relatively dry conditions. Although it is a tolerant species, it will not do well in waterlogged conditions.[107] Rosemary will not self-seed and as such is rarely, if ever, seen outside of a cultivated setting in the UK.

Associated insect benefits

Rosemary is attractive to insects and bees in particular. The flowers are accessible to both long and short tongued bees. Bumble bees of all types forage on lavender; solitary bees tend to be of the larger types, including some flower bees which are uncommon to absent in some areas of the UK. A relative newcomer to the UK, the violet carpenter bee (*Xylocopa violacea*) has been sighted in scattered location, with several observations being from rosemary.[1]

In the right climate, rosemary can yield a significant honey crop; as much as 200 kg/Ha is cited. The honey is golden and of very fine flavour. Narbonne honey—a speciality of the Languedoc region of south-western France, is said to bear nuances of flavour from the forage species, thought to be rosemary. The honey is a pale blond granulated honey with a fine grain structure.[5]

Bee forage potential[1]		Notable foraging bee species[1]
Nectar/pollen	N/P	Honey bee, *Apis mellifera*
Honey bees	✿✿	Common carder bee, *Bombus pascuorum*
Short-tongued bumble bees	✿✿	Hairy-footed flower bee, *Anthrophora plumipes*
Long-tongued bumble bees	✿✿	Violet carpenter bee, *Xylocopa violacea*
Solitary bees	✿✿	

Pollen presentation

Pollen loads from rosemary are mauve-purple.[108] The pollen grains are ~30 μm. Like other related species in the lamiaceae family, they have six pronounced colpate apertures that can give the grain an almost hexagonal appearance. The exine is thin-medium with a pitted surface texture.[109]

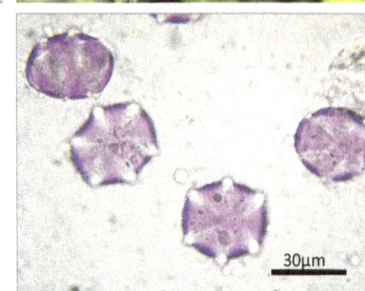

30μm

Aquifoliaceae—Holly

Aquifoliaceae is a family containing a single genus of 570 species—that of *Ilex* or holly. The genus includes the greatest number of any woody, dioecious angiosperm genus. All are evergreen or deciduous trees, shrubs or climbers spanning the tropics through to temperate zones. The majority are slow-growing with alternate glossy leaves, frequently with a spiny leaf margin and inconspicuous flowers.

Family	Aquifoliaceae
Species	*Ilex aquifolium*
Common name	Holly
Flowering time	May-June

Species notes

Holly, *Ilex aquifolium* is a UK-native evergreen tree which is increasingly common in deciduous woodland. Mature specimens can reach 15 m. It is often found as part of hedgerows, in both urban and rural settings. Leaves are alternate, dark green and glossy.[25] Lower leaves are frequently characteristically spiny although upper leaves are frequently non-spiny. This variation in leaf form on the same plant, known as heterophylly, has been shown in holly to be a response to herbivore activity.[110]

Flowers are small and four-petalled. Male flowers are creamy, sometimes tinged with pink and borne in axillary groups (A,B). Each flower has four prominent stamens. Female flowers are in much smaller groups creamy to white in colour, and have a prominent green four-lobed stigma on top of the ovary (C). Holly is entomophilous; the resulting fruit is a red drupe (D).[1]

Habitat

Holly is a pioneer species, and will tolerate most environments, preferring moist shady conditions in forests or shady slopes. It will quickly colonise forest habitats, and is considered an invasive species in some territories.[111,112]

Associated insect benefits

Holly flowers secrete nectar freely and are attractive to insects—the open flowers are accessible to all types of bees. The flowering period is relatively short, however (2-3 weeks).[1] Holly leaves are eaten by the larvae of a number of Lepidoptera species, including the holly blue butterfly (*Celastrina argiolus*), yellow-barred brindle (*Acasis viretata*), double striped pug (*Gymnoscelis rufifasciata*) and holly tortrix (*Rhopobota naevana*).[113]

Bee forage potential[1,49]		Notable foraging bee species[1]
Nectar/pollen	N/P	Honey bee, *Apis mellifera*
Honey bees	✿ ✿	Buff-tailed bumble bee, *Bombus terrestris*
Short-tongued bumble bees	✿ ✿	White-tailed bumble bee, *Bombus leucorum*
Long-tongued bumble bees	✿	Mining bees—*Andrena* spp.
Solitary bees	✿	

Pollen presentation

Pollen loads are yellow-green.[11] Pollen grains are ~25 μm in diameter, round in cross section with a deeply netted surface structure. There are three colporate apertures. The exine is medium to thick, with very visible coarse rods.[10,114]

Holly has a pollen coefficient of 50 and is normally represented in honey.[3]

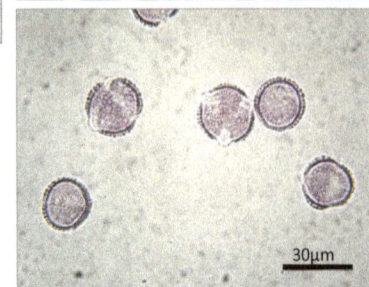

Asteraceae—Dandelion

Asteraceae is a substantial family of flowering plants including over 32,000 species across >1900 genera. Most are annual, biennial or perennial plants; a small proportion are shrubs or trees. The family is represented across all territories globally with the exception of Antarctica. Many species exhibit flower heads that are composed of many (sometimes hundreds) of individual florets surrounded by protective bracts.

Family	Asteraceae
Species	*Taraxacum officinale* and others
Common name	Dandelion
Flowering time	March-October peaking in April-May

Species notes

Dandelion, *Taraxacum officinale*, is an abundant perennial—a complex of 232 microspecies many of which require expert knowledge to tell apart. The plant will reach ~20 cm. Hairless to sparsely hairy leaves form in a basal rosette; leaves can be deeply or shallowly toothed depending upon maturity and type. The leaf and flower stems will bleed a white gummy sap when broken. Large solitary yellow flower heads, 20-60 mm across rise on hollow, leafless stems that are often tinged with red. Flowers only open in full sun (B,C). After flowering, a halo of feathery seeds forms, following structural changes in the flower head (D). These are readily dispersed on the wind.[25]

Habitat

Dandelion is ubiquitous in the UK, and will quickly colonise any grassy or rough ground, irrespective of soil type, particularly after soil disturbance.[25] A deep tap root makes the plant resilient to dry conditions.

Associated insect benefits

Dandelion is one of the most valuable of wild plants for pollinators. It is a major source of nectar and pollen, and its plentiful flowering early in the season provides key spring resources.[1] The flower tubes vary from 3-7 mm, and are accessible to all types of bees as well as early flies and leaf or pollen beetles.[36] The closing of the flowers in poor weather protects the nectar and pollen reserves from rain.[1] Dandelion pollen is particularly high in lipids,[115] although has a crude protein content of 13.2-22.7%, which is relatively poor in terms of nutritional quality.[2] It will yield copious nectar from which honey bees may readily gather a surplus of intense and sharply flavoured golden yellow honey which will granulate quickly with a coarse grain structure. Dandelion may produce up to 200 kg/Ha under optimum conditions.[5]

Bee forage potential[1]		Notable foraging bee species[1]
Nectar/pollen	N/P	All common species as well as a wide
Honey bees	🐝🐝🐝	range of solitary bees including *Andrena*,
Short-tongued bumble bees	🐝🐝🐝	*Colletes* and *Lasioglossum* spp.
Long-tongued bumble bees	🐝🐝🐝	Mason bees, *Osmia* spp.
Solitary bees	🐝🐝🐝	Small stem-nesting bees, *Hylaeus* spp.

Pollen presentation

Pollen loads from dandelion are orange.[11] Pollen grains are ~25 µm in diameter, with three colporate apertures. The exine is characteristically structured with spines and infoldings.[116] Dandelion has a pollen coefficient of 10 and is under-represented in honey.[117]

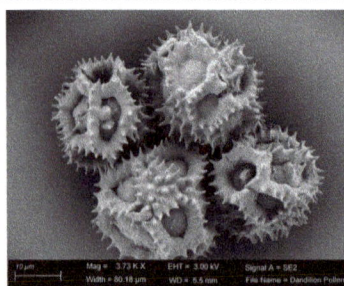

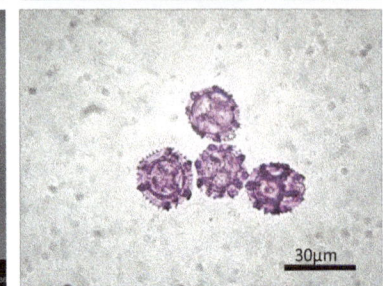

30µm

Asteraceae—Dandelion

Asteraceae—Ragwort

Family	Asteraceae
Species	*Senecio jacobaea*
Common name	Ragwort
Flowering time	June-October, peaking in July

Species notes

Common ragwort, *Senecio jacobaea*, is a ubiquitous perennial reaching 30-150 cm in height. The plant bears characteristically deeply and intricately lobed leaves (A,B). Stems are largely hairless, whereas the leaves can have sparse, cobwebby hairs on the underside. Basal leaves are usually in a rosette, which can die off by flowering time (A). Yellow flowers are borne in dense clusters, each flower head being ~15-25 mm across with ray florets ~5-9 mm long (C). Closely related to Oxford ragwort (*Senecio squalidus*), hoary ragwort (*Senecio erucifolius*) and marsh ragwort (*Senecio aquiaticus*), which can be differentiated based on leaf and flower morphology.[25]

Habitat

Ragwort is common in grassy places and waste ground particularly, especially on light, medium or calcareous soils.[1,25]

Associated insect benefits

Ragwort is among the most visited of species by a range of bee species, more so by bumble bees than honey bees though. The open flowers are accessible to both long- and short-tongued species. It is also highly valued by solitary bees, including a range of *Andrena, Colletes* and *Lasioglossum* species; leafcutter bees (*Megachile* spp.) which visit for both pollen and nectar.[1] The crude protein content of the pollen is only 16.8% however, which is nutritionally poor on average.[2]

Ragwort is notably attractive to hoverflies, in particular the long hoverfly (*Sphaerophoria scripta*), marmalade hoverfly (*Episyrphus balteatus*), hairy-eyed flower fly (*Syrphus torvus*), and Lapland syrphid fly (*Lapposyrphus lapponicus*), as well as *Syrphys vitripennis,* and *Syrphus ribesii*.[33]

Ragwort is often seen covered with orange and black striped caterpillars—the larvae of the cinnabar moth, *Tyria jacobaeae,* which feed preferentially on *Senecio* species (D).[118]

Although ragwort yields ample nectar, it is not renowned as a honey cropping species other than in drought years when there may be a shortage of forage from other sources. The honey is deep yellow in colour and is said to have an unpleasant flavour, though this may ameliorate with keeping.[1]

Bee forage potential[1]		Notable foraging bee species[1]
Nectar/pollen	N/P	Red-tailed bumble bee, *Bombus lapidarius*
Honey bees	✿	Mining bees, *Colletes daviesanus; C. fodiens*
Short-tongued bumble bees	✿ ✿	Hairy-legged mining bee, *Dasypoda hirtipes*
Long-tongued bumble bees	✿	Cuckoo bees, *Epeolus, Nomada, Stelis* spp.
Solitary bees	✿ ✿ ✿	

Pollen presentation

Pollen loads from ragwort are orange.[119] The pollen grains are ~25-30 μm in diameter, with three sunken colporate apertures. The exine is medium-thick, ornamented with spines.[120] Pollen morphology highly characteristic of the asteraceae family.

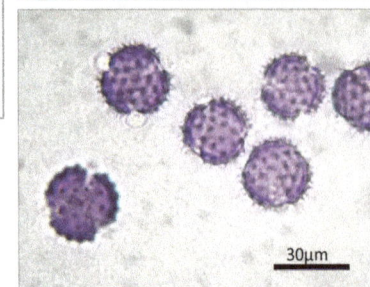

30μm

84

Asteraceae—Ragwort

Asteraceae—Sunflower

Family	Asteraceae
Species	*Helianthus* spp.
Common name	Sunflower
Flowering time	August-October depending on variety

Species notes

A number of **sunflower** species exist, many as decorative cultivars. The annual common sunflower, ***Helianthus annuus*** is grown both as a garden flower, and also commercially for the seed for its oil content and as bird– and whole-food uses. It is a tall, fast-growing annual reaching typically 3 m, with broad, heart-shaped leaves and coarsely hairy stems (B,C). Flowers up to 30 cm across are borne at the top of the stem. *Helianthus annuus* can produce multiple flower heads.

Habitat

A native of North America, sunflower was brought to Europe in the 16th century. This species grows best in fertile, moist and well-drained soils.

Associated insect benefits

Honey bees are considered an essential pollinator for commercial sunflower crops, partly because of flower constancy. It is thought that sensory cues such as pollen odour, diameter of seed head and height may influence the visitation of honey bees to sunflower.[121]

Although a common species from which to obtain honey crop in mainland Europe, sunflower is rarely grown in adequate quantities for this to be the case in the UK. The honey potential is relatively low, at 26-50 kg/Ha and the sugar value low (0.12-0.3 mg sugar/flower/24 h). When honey is obtained, it is variously described as egg-yolk-yellow, with a strong aroma and characteristic flavour, dark with unpleasant flavour, or mild and quick to granulate, the variation perhaps depending upon environmental factors.[1,5]

Bees of all types as well as other pollinating insects visit the flower heads for both nectar and pollen. Pollen is prolific across the broad flower head. However its crude protein content can be poor (13.8-30.6%),[2] though these data are given for *Helianthus* spp. and there may be differences between individual species. Many species of Lepidoptera, Coleoptera and some Hymenoptera are attracted to the showy flowers.

Bee forage potential[1]		Notable foraging bee species[1]
Nectar/pollen	N/P	Honey bee, *Apis mellifera*
Honey bees	✿ ✿	
Short-tongued bumble bees	✿ ✿	Bumble bees and solitary bees of many species.
Long-tongued bumble bees	✿ ✿	
Solitary bees	✿ ✿	

Pollen presentation

Pollen loads from ragwort are yellow.[11] The pollen grains are ~25-30 µm in diameter, with three sunken colporate apertures. The exine is thin, ornamented with spines.[122]
Sunflower has a pollen coefficient of 10, and is under-represented in honey.[3]

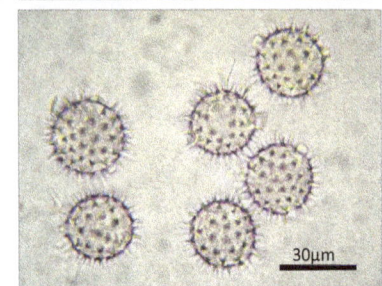

30µm

Asteraceae—Knapweed

Family	Asteraceae
Species	*Centaurea nigra*
Common name	Common knapweed
Flowering time	June-September

Species notes

Common knapweed, *Centaurea nigra* is a common perennial thistle-like plant reaching 20-80 cm. It has an erect growth habit, with oppositely branched hairy stems (A). Leaves are also hairy, stalked basal leaved are usually entire, whereas stalkless upper leaves can be lobed. Purple flowers ~15-20 mm across are borne at the top of the stems with bracts forming a globular head below the florets (B). Florets are usually all tubular, but the outer florets can be long and forked giving a fringed appearance.[25]

Habitat

Knapweed is common across all types of rough grassland, particularly on damp or heavy soils, field borders and old meadows.[25]

Associated insect benefits

This species in very rich in nectar and therefore a valuable forage source to insects of different types. Among its predominant Hymenopteran visitors is *Bombus lapidarius,* the red-tailed bumble bee and *Bombus pascuorum,* the common carder bee. There are also a number of solitary bee visitors that have a known preference for knapweed, including *Andrena marginata*, an uncommon mining bee mainly found in southern England and Wales. The large scabious bee (*Andrena hattorfiana),* collects pollen from knapweed as well as scabious—this is an attractive mining bee that is listed as rare in the UK and numbers are thought to be declining. It is found principally in chalk grassland areas. The nomad cuckoo bee (*Nomada armata*) lays its eggs in the nests of *A. hattorfiana,* and so is also threatened.

In parts of Ireland, honey bees can gather a surplus from knapweed[1]—the honey is a light amber in colour and thin, with a sharp flavour and slightly bitter after taste. It crystallises with a soft granulation.[5]

Bee forage potential[1]		Notable foraging bee species[1]
Nectar/pollen	N/P	Honey bee, *Apis mellifera*
Honey bees	✿ ✿	Red-tailed bumble bee, *Bombus lapidarius*
Short-tongued bumble bees	✿ ✿ ✿	Common carder bee, *Bombus pascuorum*
Long-tongued bumble bees	✿ ✿ ✿	Mining bee, *Andrena marginata*
Solitary bees	✿ ✿ ✿	Large scabious bee *Andrena hattorfiana*

Pollen presentation

Pollen loads from knapweed are grey-white.[11] The pollen grains are ~35-40 µm in diameter, with three colporate apertures. The exine is medium-thick, with spaced rods and ornamented with spines.[123]

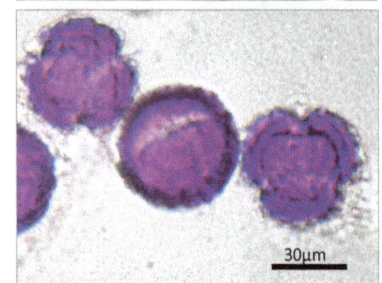

30µm

Asteraceae—Sea Aster

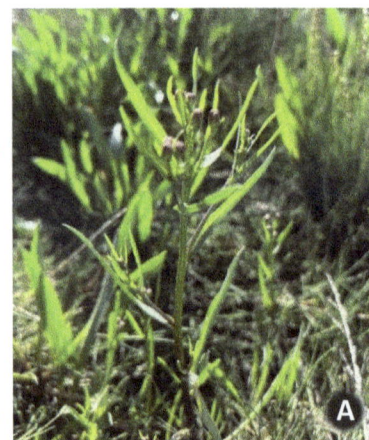

Family	Asteraceae
Species	*Aster tripolium, (previously Tripolium pannonicum)*
Common name	Sea aster
Flowering time	July-October

Species notes

Sea aster is a biennial plant with multiple branching stems reaching ~15-50 cm and often forming dense clumps. fleshy leaves, basal leaves being lanceolate and stalked, stem leaves more strap-like and stalkless. Two varieties occur in the UK—**A. tripolium flosculosus**, the dominant form in East Anglia, has flower heads ~10 mm across with yellow disk florets only. **A. Tripolium tripolium** has a yellow centre of disk florets surrounded by purple ray florets, and are ~15-20 mm across, looking not unlike a Michaelmas daisy (B,C). After pollination, the flowers produce downy seeds with a plume of brown hairs.[25]

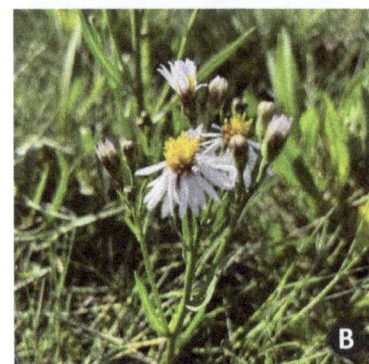

Habitat

Sea aster is native to Eurasia and North Africa. This species is a notable halophyte—a salt tolerant plant that can tolerate, and indeed, prefers areas of high salinity. Hence it is found exclusively in coastal areas of salt marshes, tidal rivers and other brackish environments (D).[1] The sea aster can help to stabilise areas of mud and sand, reducing coastal erosion and forming habitats for less salt-tolerant species to colonise.[124]

Associated insect benefits

The plant flowers well into autumn and hence provides a valuable source of nectar for late-flying butterflies such as painted lady (*Vanessa cardui*), and red admiral (*Vanessa atalanta*).[1] Honey bees will forage from this species, and Kirk notes that bee hives are sometimes moved to coastal sites abundant in sea aster to take advantage of the opportunity for a honey crop.[1] Honey is light to medium amber in colour and with a characteristic and slightly salty flavour.[5] The sea aster is also visited particularly by the mining bee (*Colletes halophilus*), which occurs mainly along southern and eastern coasts of the UK.[1]

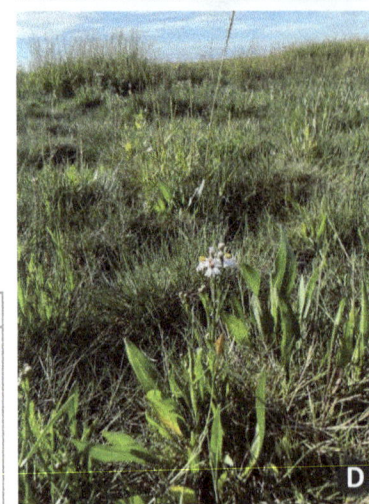

Bee forage potential[1]		Notable foraging bee species[1]
Nectar/pollen	N/P	Honey bee, *Apis mellifera*
Honey bees	✿✿✿	Common carder bee, *Bombus pascuorum*
Short-tongued bumble bees	✿✿	Mining bee species, *Colletes halophilus*
Long-tongued bumble bees	✿✿	and various *Lasioglossum* spp.
Solitary bees	✿✿✿	

Pollen presentation

Pollen loads from sea aster are yellow-orange.[9] The pollen grains are ~25-30 μm in diameter, with three colporate apertures. The exine is medium-thick and the dotted surface texture shows its clear ornamentation with spines.[9,125,126]

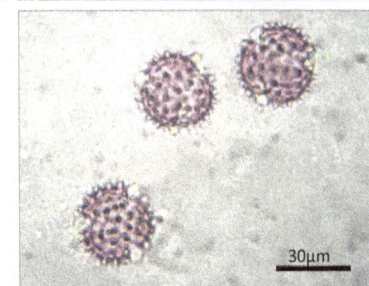

30μm

Araliaceae—Ivy

Araliaceae is a family of flowering plants including 43 genera and around 1500 species, primarily woody plants and some herbaceous species. Morphology among the family varies widely and there is no unifying characteristic, other than a woody habit and flowering habit in simple umbels. Although Araliaceae species occur predominantly in the tropics some are native to temperate regions. Many are used as ornamentals.

Family	Araliaceae
Species	*Hedera helix*
Common name	Common Ivy
Flowering time	September-November

Species notes

Common ivy, *Hedera helix*, is a common evergreen climber which supports itself with numerous short roots from the stems. A mature ivy can achieve substantial size, with shoots exceeding 30 m, easily covering a sizeable tree. Leaves are glossy dark green, 4-10 cm long sometimes with a pale marbling. Those on the creeping or climbing stems are usually 3- or 5-lobed (see pressing). Leaves on flowering stems are unlobed, diamond shaped (A). Flowers are borne in umbels (B). Individual flowers have five tiny sepals, five yellowish green petals and five prominent stamens.[25] The fruit are purple-black berries 6–8 mm in diameter, ripening in late winter (D); each contains up to five seeds.

Habitat

Common ivy is native to Europe and western Asia. It prefers moist, shady locations, and will thrive in a wide range of soil types. Its ability to climb varies with support-plant or substrate variety, though typically, it will prefer a non-reflective rough surface with near-neutral pH.

Associated insect benefits

Ivy is a key source of late-season forage for insects of all types. The flowers are open and secrete so freely that they can literally drip with nectar.[1] Honey bees will easily gather a surplus from ivy, though so late in the season that it is rarely harvested, moreover, the flavour is said to be unpleasant, although it will mellow with age. The honey potential is high, estimated at up to 500 kg/Ha.[5]

Over 30 species of hoverfly have been observed feeding on ivy including the batman hoverfly (*Myathropa florea*) and the hornet mimic hoverfly (*Volucella zonaria*).[127] Also wasps are drawn to the sugary nectar,[36] the common wasp *Vespula vulgaris* being most prevalent. The ivy bee (*Colletes hederae*), a solitary bee that first arrived in the UK in 2001, emerges with the flowering of ivy, its main pollen and nectar source. Ivy is key to this species' survival.

Late season butterflies such as red admiral (*Vanessa atalanta*) can also be seen foraging on ivy.[127]

Bee forage potential[1]		Notable foraging bee species[1]
Nectar/pollen	N/P	Honey bee, *Apis mellifera*
Honey bees	✿✿	Common carder bee, *Bombus pascuorum*
Short-tongued bumble bees	✿	Ivy bee, *Colletes hederae*
Long-tongued bumble bees	-	
Solitary bees	✿✿✿	

Pollen presentation

Pollen loads from ivy are yellow-green.[11] Pollen grains are ~25-30 μm in diameter, round but slightly triangular cross section with reticulate surface structure. The exine is of medium thickness with spaced rods. There are three colporate apertures.[128]

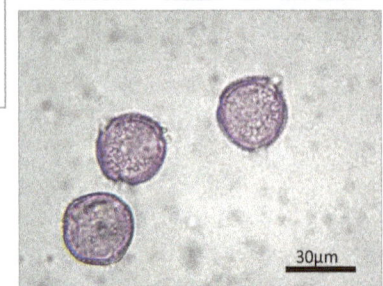

30μm

Apiaceae—Carrot

Apiaceae is a family of mostly aromatic flowering plants including over 3800 species in ~446 genera, principally in the temperate northern hemisphere. It is characterised by an alternate leaf arrangement and leaved that are often deeply lobed or dissected. Flowers are nearly always clustered in terminal umbels with hermaphroditic blossoms bearing five petals and five stamens; petals may be of unequal size. Small ridged fruits are produced on maturity.[129]

Family	Apiaceae (Umbelliferae)
Species	*Daucus carota*
Common name	Wild carrot, Queen Anne's lace
Flowering time	June-August

Species notes

Wild carrot, *Daucus carota*, is a tall, roughly hairy biennial plant, with robust ridged stems reaching 1 m in height. Leaves are repeatedly divided into deeply lobed leaflets (A). The white flowers are in umbels 3-7 cm across; often there is a single deep red flower in the centre (B). A ring of finely divided bracts below the flower heads are characteristic of this species (C), as is the balling tendency of the seed heads (D). The fruits are ovoid and flattened with long, hooked spines that facilitate their transport in animal fur.[25]

Habitat

A native to the UK, wild carrot is common on rough grassland, scrub and verges, particularly over dry calcareous soils and in coastal areas. However it is increasingly being used as part of 'wild flower' seed mixes, and can be seen on grassy verges in areas where it would not otherwise be usual.[25]

Associated insect benefits

The nectar of this species is exposed, and so accessible to many short-tongued insects.[1] It is particularly attractive to flies—in a study published by Klecka and colleagues, *Daucus carota* was seen to have the greatest diversity of hoverfly visitor species of the 57 plant species covered in the study.[33] Honey bees and bumble bees have been seen to visit, but not so frequently as solitary bees across a number of different families. One of the mining bees, *Andrena nitidiuscula*, is dependent upon wild carrot and closely related plants for pollen. It is a small black bee found only in southern England.[1]

Honey has occasionally been taken from carrot when it has been planted extensively for seed in the US. The honey was a light amber colour.[1,11]

Bee forage potential[1]		Notable foraging bee species[1]
Nectar/pollen	N/P	Mainly solitary bees incl. *Andrena, Colletes,*
Honey bees	⊛	*Lasioglossum* spp.
Short-tongued bumble bees	⊛	Leafcutter bees, *Megachile* spp.
Long-tongued bumble bees	⊛	Cuckoo bees, *Epeolus, Nomada, Sphecodes* spp.
Solitary bees	⊛ ⊛ ⊛	*Andrena nitidiuscula*

Pollen presentation

Pollen loads are grey-white.[130] Pollen grains are longitudinal oval (prolate), characteristic of Apiaceae ~15x30 µm, with smooth surface structure and thin exine. There are three colporate apertures.[131]

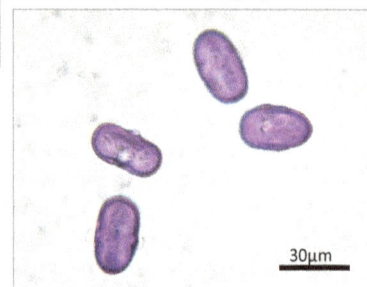

Iridaceae—Crocus

Iridaceae is a family of monocotyledon perennial plants including ~2500 species across 69 genera. All are perennial plants, growing from a bulb, corm or rhizome. They are characterised by erect growth and long grass-like leaves, often with a central fold. The sub-family Crocoideae, which includes the genus *Crocus* is one of the major subfamilies of Iridaceae. Most are from Africa but some members are native to Europe and Asia.

Family	Iridaceae
Species	*Crocus* spp
Common name	Crocus, spring crocus
Flowering time	February-March

Species notes

Crocus refers to a genus of seasonal flowering plants which comprises about 100 species. Most common in the UK are *Crocus sativus*, the autumn-flowering or saffron crocus (a cultivar, unknown in the wild),[132] and ***Crocus vernus****,* the spring crocus, which is non native, but naturalised. The latter is native to the Alps, Pyrenees and Balkans, and cultivated widely as an ornamental. Growing from an underground corm, upright flowers emerge before the narrow leaves. Flowers are white through to dark purple (A-C), with a triple-branched stigma and three anthers. The yellow crocus, (D) *C. x stellaris*, has yellow flowers and slightly narrower leaves.[25]

Habitat

Crocus is largely found as a planted cultivar, although there is some naturalisation in the south of England.[25]

Associated insect benefits

Crocuses are a key source of pollen early in the year when little else is available for insects. The flowering period is long, and the flower offers both nectar and pollen to visiting foragers. Nectar is secreted at the base of the flower tube, only readily accessible by longer tongued species. However if the nectar accumulates, rising up the flower tube, it becomes reachable (with some effort!) by shorter tongued bumble bees and honey bees. Pollen is easily accessible from the prominent anthers, and is abundant at a time which is key for colony growth in honey bees and establishment for bumble bees—it is not unusual to see bumble bee queens working the crocus in spring.[1]

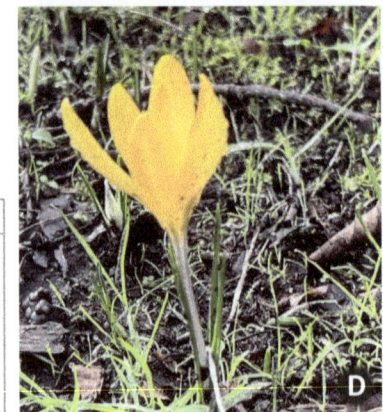

Bee forage potential[1]		Notable foraging bee species[1]
Nectar/pollen	N/P	Honey bee, *Apis mellifera*
Honey bees	✿✿✿	Buff-tailed bumble bee, *Bombus terrestris*
Short-tongued bumble bees	✿	Red-tailed bumble bee, *Bombus lapidarius*
Long-tongued bumble bees	-	Early bumble bee, *Bombus pratorum*
Solitary bees	✿	Tree bumble bee, *Bombus hypnorum*

Pollen presentation

Pollen loads are orange.[11] Pollen grains are notably large, >100 μm but variable in size, round. The exine is thin, but there is a very thick intine, giving the grains a characteristic appearance under the light microscope. Small dots may be seen, indicating very small surface spines. There are no apertures.[3,133]

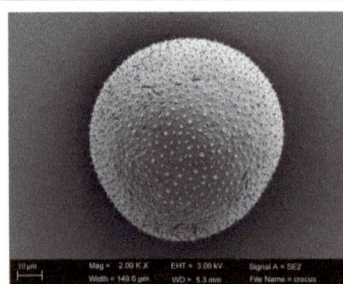

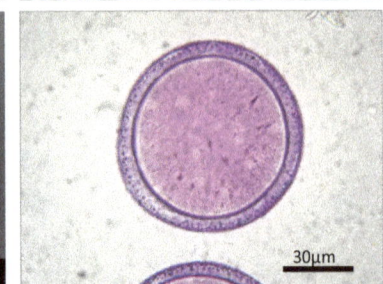

30μm

Amaryllidaceae—Chives

Amaryllidaceae is a family of monocotyledon herbaceous, mainly perennial flowering plants, the majority growing from bulbs. It is characterised by linear, sometimes strappy leaves and flowers which are arranged in umbels on the stem. The family includes ~1600 species divided into 70-75 genera, among them many ornamentals and vegetables, including daffodils, snowdrops, chives, leeks and garlic.

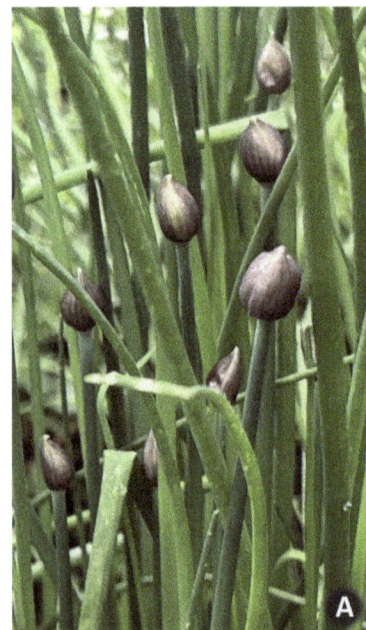

Family	Amaryllidaceae
Species	*Allium schoenoprasum*
Common name	Chives
Flowering time	July-September

Species notes

Chive, *Allium schoenoprasum*, is a bulb-forming herbaceous plant growing to ~30-40 cm, forming dense clusters from the roots. The leaves and stems are hollow and tubular, just a few mm across (A), and with a soft, slightly squeaky texture. Pale purple flowers, each a dense inflorescence of 10-30 flowers are borne at the top of the flower stems (B). Each individual flower has six petals, and six stamens (C). Seeds are produced in a three-valved capsule following pollination. As with all alliums, chives contain the enzyme alliinase, which produces volatile sulphur compounds if the plant tissue is damaged, resulting in their strong aroma.[134]

Habitat

Chives are native to temperate areas of the northern hemisphere.[135] They thrive in well-drained soil, preferably rich in organic matter, a pH of 6-7 and in full sun.

Associated insect benefits

The volatiles from chive mean that in general, it is unattractive to insects. However the flowers do attract bees, both honey bees and bumble bees, for nectar and pollen.[1] The flowers produce a small amount of nectar of reasonable quality (0.37-0.48 mg sugar per flower per 24 h) and a honey crop is not unknown from areas where chive is cultivated. The honey is a light amber colour, and is said to have an 'oniony' aroma which will ameliorate with storage.[5]

Chives are also a food plant for the larvae of the leek moth or onion leaf miner, *Acrolepiopsis assectella*, a small moth considered to be an invasive species in North America.[136]

Bee forage potential[1]		Notable foraging bee species[1]
Nectar/pollen	N/P	Honey bee, *Apis mellifera*
Honey bees	✿ ✿	Red-tailed bumble bee, *Bombus lapidarius*
Short-tongued bumble bees	✿ ✿	Common carder bee, *Bombus pascuorum*
Long-tongued bumble bees	✿ ✿	
Solitary bees	✿	

Pollen presentation

Pollen on the stamens of *Allium schoenoprasum* appears grey-white. Pollen grains are ovoid semi-circular, ~20-30 µm along the longest axis. The exine is thin and the surface structure smooth, and under light microscopy, lacking in other features. The pollen grains have has a single sulcate aperture.[137]

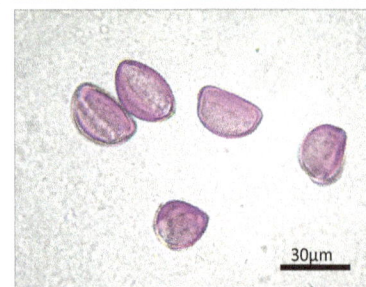

30µm

Amaryllidaceae—Snowdrop

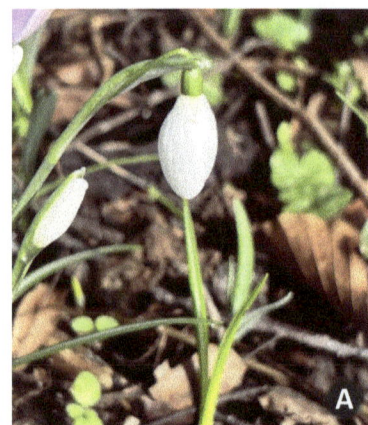

Family	Amaryllidaceae
Species	*Galanthus nivalis*
Common name	Snowdrop
Flowering time	January-March

Species notes

Snowdrop, *Galanthus nivalis*, is a small perennial plant, reaching 15-20 cm, although variable depending upon variety. Growing from a bulb, the strap-like leaves are flat, uniformly green and ~1 cm across. Each bulb produces a solitary pendulous flower whose flower stalk is cowled by a green, leaf-like bract. Flowers are ~12-25 mm, with three outer sepals and three inner petals that are tipped green forming a cup that houses six stamens (A,B), There are many cultivars including some ornamental double-petalled varieties.[25]

Habitat

Although not native to the UK, snowdrop is widely naturalised. It was brought over in the sixteenth century as a garden plant, and first recorded in the wild in 1778. Snowdrop favours damp woodland and other shady places with soils rich in organic matter.[25]

Associated insect benefits

Snowdrop is among the first of the spring flowers to make an appearance, and can remain in bloom for several weeks. Although too early for any bumble bee or solitary bee species, it is a favourite of honey bees, who seek out the plentiful pollen. The flowers do secrete a small quantity of nectar, which forms in depressions on the inner sides of the petal and the base of the flowers. Kirk notes that honey bees adopt a characteristic posture when foraging on snowdrop, inserting their head and forelegs and middle legs into the flower, and grasping the flower with the hind legs (C). From there she can brush the anthers with the forelegs and deposit pollen into the corbiculae on the hind legs.[1]

Bee forage potential[1]		Notable foraging bee species[1]
Nectar/pollen	n/P	Honey bee, *Apis mellifera*
Honey bees	✿ ✿	
Short-tongued bumble bees	-	
Long-tongued bumble bees	-	
Solitary bees	-	

Pollen presentation

Pollen loads from *Galanthus nivalis* are orange.[11] Pollen grains are ovoid semi-circular, typical in shape of the amaryllidaceae, ~20-25 μm in along the longest axis. The exine is thin and the surface structure smooth. The pollen grains have a single sulcate aperture.[138]

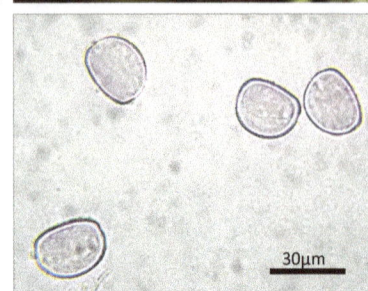

30μm

Amaryllidaceae—Snowdrop

Asparagaceae—Bluebell

Asparagaceae is a family of monocotyledon herbaceous, mainly perennial flowering plants. The family includes ~2900 species across 114 genera.[139] Most are herbaceous perennials with long simple leaves that form a tight rosette at the base; many are geophytes rooting from bulbs or corms, and often with showy flowers that have identical petals and sepals as well as stamens in multiples of three.[25]

Family	Asparagaceae
Species	*Hyacinthoides non-scripta; Hyacinthoides hispanica*
Common name	Bluebell (English/Spanish respectively)
Flowering time	April-June

Species notes

Bluebell is an iconic clump-forming fleshy flowering plant reaching 20-50 cm. Leaves are strap-like, 7-15 mm across with a shallow central crease. The UK native bluebell, *Hyacinthoides non-scripta,* has an elegant flower spike , slightly drooping at the tip, and with flowers all hanging downward to one side of the stem (A). Flowers are 14-20 mm long, and form a parallel sided tube of 6 tepals. The Spanish bluebell, *Hyacinthoides hispanica* is similar, slightly more robust, but with flower spikes not one-sided, and flowers more bell-shaped broadening toward the mouth. Fertile hybrids, closer in form to and more widely naturalised than *H. hispanica,* form readily (B), as do pink/white variants, which do occur, but are rarer in *H. non-scripta* (C).[25]

Habitat

The UK native bluebell, *Hyacinthoides non-scripta* is abundant in woodlands, shady banks and hedgerows, mostly lowland areas. The Spanish bluebell, *Hyacinthoides hispanica* has similar habitat requirements, but is more tolerant of drier, sunnier aspects.[25]

Associated insect benefits

Bluebell flowers secrete large quantities of nectar, but the long flower tubes are too long for short-tongued species. Main visitors are bumble bee queens as they have longer tongues than the workers and are able to access the nectar reserves. Shorter tongued species are able to access nectar from the side of the flower, but do so without pollinating the flower. Pollen is collected by honey bees, solitary bees including *Andrena* spp. and *Osmia* spp. The large bee fly (*Bombylius major*) is a common visitor.[1]

Bee forage potential[1]		Notable foraging bee species[1]
Nectar/pollen	N/P	Honey bee, *Apis mellifera*
Honey bees	✿	Red-tailed bumble bee, *Bombus lapidarius*
Short-tongued bumble bees	✿ ✿	Garden bumble bee, *Bombus hortorum*
Long-tongued bumble bees	✿ ✿ ✿	Common carder bee, *Bombus pascuorum*
Solitary bees	✿	

Pollen presentation

Pollen loads from bluebell are blue-green.[11] Pollen grains are ~50-70 µm along the longest axis. The grains are ovoid semi-circular with a thin exine but a medium-thickness intine; the surface is smooth. There is a single sulcate aperture.[139]

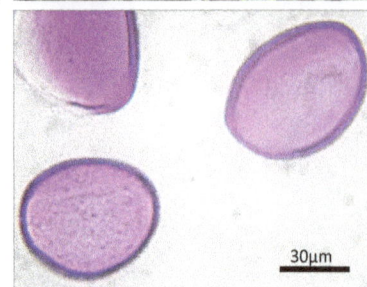

30µm

Asparagaceae—Asparagus

Family	Asparagaceae
Species	*Asparagus officinalis*
Common name	Asparagus
Flowering time	June-September

Species notes

Garden asparagus, *Asparagus officinalis*, is a tall, delicate dioecious perennial reaching 1-2 m. It has delicate, needle-like foliage—true leaves are reduced to scales on the stem, whereas the 'leaves' are flexible, thread-like stems up to 30 mm long, growing in clusters.

Small, (~3-6 mm), bell-shaped yellow flowers are borne in groups of 1-4, growing from the axils of the true leaves. The flowers (A, B) have 3 sepals, 3 petals.[25]

Female plants bear small round berries which mature through green and brown to a bright red (C).

Habitat

A native of Europe and temperate regions of Asia, asparagus has been cultivated in the UK since Roman times, and is widely naturalised in the south-east of England. It thrives on dunes, heaths and grassland, particularly on dry, sandy soils.[25]

Associated insect benefits

The flowers of asparagus are highly attractive to bumble bees and honey bees for both nectar and pollen.[1] The pollen in particular is highly valuable, offering a crude pollen content of 37%, which is among the highest of the values reported in listings given in the literature.[2]

Although asparagus is not a common honey-yielding species, honey has been obtained from cultivated asparagus, and has been described both as greenish in colour and of mediocre quality (France), and amber to dark in colour, with a low market value compared with other honeys (California).[1]

Asparagus is noted for attracting its namesake asparagus beetle, small leaf beetles of the *Crioceris* genus. Both adults and larvae of some species feed on asparagus. Two species common in the UK are *Crioceris asparagi,* a metallic blue-black beetle with a red bordered elytra and cream-coloured spots and *Crioceris duodecimpunctata*, the spotted asparagus beetle, which is bright red, with black spots on the elytra and black antennae.

Bee forage potential[1]		Notable foraging bee species[1]
Nectar/pollen	N/P	Honey bee, *Apis mellifera*
Honey bees	✿ ✿	Common carder bee, *Bombus pascuorum*
Short-tongued bumble bees	✿ ✿	Hairy-footed flower bee, *Anthrophora plumipes*
Long-tongued bumble bees	✿ ✿	Violet carpenter bee, *Xylocopa violacea*
Solitary bees	✿	

Pollen presentation

Pollen loads from asparagus are orange.[11] Pollen grains are ~20-25 µm along the longest axis. The grains are ovoid semi-circular with a thin exine; the surface is smooth. There is a single sulcate aperture.[140]

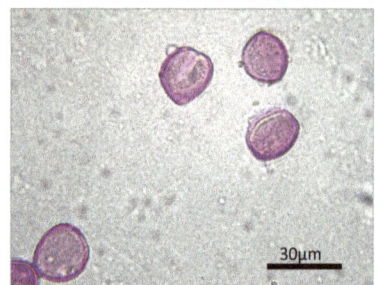

30µm

References

1. Kirk, W.D.J., & Howes, F.N. (2012). *Plants for Bees*. International Bee Research Association. ISBN 978-0-86098-271-5.

2. Aston, D., & Bucknall, S. (2021). *Good Nutrition, Good Bees*. Northern Bee Books. ISBN 978-1-912271-95-5.

3. Sawyer, R. (1988). *Honey Identification*. Cardiff Academic Press. ISBN 978-1-904846-53-6.

4. Halbritter, H., Ulrich, S., Grímsson, F., Weber, M., Zetter, R., Hesse, M., Buchner, R., Svojtka, M., & Frosch-Radivo, A. (2018). *Illustrated Pollen Terminology*. Springer. ISBN 978-3-319-71365-6.

5. Crane, E. (1979). *Honey: A Comprehensive Survey*. Heinemann. ISBN 434-90270-5.

6. Christenhusz, M., & Byng, J. (2016). The number of known plant species in the world and its annual increase. *Phytotaxa*, 261, 201-217.

7. Britannica, The Editors of Encyclopaedia. (2017). *Papaveraceae. Encyclopedia Britannica*. Retrieved from https://www.britannica.com/plant/Papaveraceae. Accessed 9 October 2023.

8. Blamey, M., & Grey-Wilson, C. (1989). *The Illustrated Flora of Britain and Northern Europe*. Hodder & Stoughton. ISBN 0-340-40170-2.

9. Roberts, S.A. (2022). *Plants of Significance to Bees*. Northern Bee Books. ISBN 978-1-914934-47-6.

10. Sawyer, R. (1981). *Pollen Identification for Beekeepers*. University College Press, Cardiff. ISBN 978-1-904-84606-2.

11. Hodges, D. (1952). *The Pollen Loads of the Honey Bee*. International Bee Research Association. ISBN 0-86098-140-1.

12. Oberschneider, W., & Heigl, H. (2020). *Papaver rhoeas*. In PalDat - A palynological database. Retrieved from https://www.paldat.org/pub/Papaver_rhoeas/303904. Accessed June 14, 2023.

13. Britannica, The Editors of Encyclopaedia. (2017). *Ranunculaceae. Encyclopedia Britannica*. Retrieved from https://www.britannica.com/plant/Ranunculaceae. Accessed 9 October 2023.

14. Oberschneider, W., & Heigl, H. (2020). *Clematis vitalba*. In PalDat - A palynological database. Retrieved from https://www.paldat.org/pub/Clematis_vitalba/303872. Accessed October 9, 2023.

15. Britannica, The Editors of Encyclopaedia. (2017). *Grossulariaceae. Encyclopedia Britannica*. Retrieved from https://www.britannica.com/plant/Ranunculaceae. Accessed 9 October 2023.

16. Heigl, H. (2020). *Ribes nigrum*. In PalDat - A palynological database. Retrieved from https://www.paldat.org/pub/Ribes_nigrum/304001. Accessed March 23, 2024.

17. Britannica, The Editors of Encyclopaedia. (2023, November 1). *Gooseberry. Encyclopedia Britannica*. Retrieved from https://www.britannica.com/plant/gooseberry. Accessed 6 November 2023

18. Janz, N., Nylin, S., & Wedell, N. (1994). Host Plant Utilization in the Comma Butterfly: Sources of Variation and Evolutionary Implications. *Oecologia*, 99(1/2), 132–140.

19. Baynes, T. S. (Ed.). (1879). *Gooseberry*. In The Encyclopædia Britannica: A Dictionary of Arts, Sciences, and General Literature (Vol. 10, p. 779). C. Scribner's sons.

20. Halbritter, H., Heigl, H., & Schneider, H. (2020). *Ribes uva-crispa*. In PalDat - A palynological database. Retrieved from https://www.paldat.org/pub/Ribes_uva-crispa/304002. Accessed March 23, 2024.

21. Petruzzello, M. (2023, June 29). List *of Plants in the Family Fabaceae. Encyclopedia Britannica*. Retrieved November 6, 2023, from https://www.britannica.com/topic/list-of-plants-in-the-family-Fabaceae-2021803.

22. Britannica, The Editors of Encyclopaedia. (2023). *Broad Bean. Encyclopedia Britannica*. Retrieved from https://www.britannica.com/plant/broad-bean. Accessed 6 November 2023.

23. Heigl, H. (2021). *Vicia faba*. In PalDat - A palynological database. Retrieved from https://www.paldat.org/pub/Vicia_faba/306048. Accessed November 6, 2023.

24. The Editors of Encyclopaedia Britannica. (2023) *Clover. Encyclopedia Britannica*, Retrieved from https://www.britannica.com/plant/clover-plant. Accessed 6 November 2023.

25. Harrap, S. (2013) *Harrap's Wild Flowers: A Field Guide to the Wild Flowers of Britain and Ireland*. Bloomsbury. ISBN 978-1-4729-6648-3.

26. Halbritter, H., Auer, W. (2021). *Trifolium pratense*. In: PalDat - A palynological database. Retrieved from https://www.paldat.org/pub/Trifolium_pratense/306432; Accessed 6 November 2023.

27. Halbritter, H., Heigl, H., Auer, W., Schneider, H. (2021). *Trifolium repens*. In: PalDat - A palynological database. Retrieved from https://www.paldat.org/pub/Trifolium_repens/306197; Accessed 23 March 2024.

28. Soil Association. (n.d.). *Benefits of Holy Hay (Sainfoin)*. Retrieved from https://www.soilassociation.org/farmers-growers/low-input-farming-advice/herbal-leys/benefits-of-holy-hay-sainfoin/ Accessed 23 March 2024.

29. Montana State University Agricultural Research Center. (n.d.). *Sainfoin*. Retrieved from https://agresearch.montana.edu/wtarc/producerinfo/agronomy-nutrient-management/Sainfoin/ Accessed 23 March 2024.

30. Pellett, F.C. (1947). *American Honey Plants*, 4th ed. New York: Orange Judd Publ. Co., Inc.

31. Halbritter, H., Heigl, H., Auer, W. (2021). *Onobrychis viciifolia*. In: PalDat - A palynological database. Retrieved from https://www.paldat.org/pub/Onobrychis_viciifolia/306131; Accessed 23 March 2024.

32. Thomas, C.D., Glen, S.W.T., Lewis, O.T., Hill, J.K., & Blakeley, D.S. (1999, February 1). Population differentiation and conservation of endemic races: the butterfly, *Plebejus argus. Animal Conservation*, 2 (1), 15–21.

References

33. Klecka, J., Hadrava, J., Biella, P., & Akter, A. (2018). Flower visitation by hoverflies (Diptera: *Syrphidae*) in a temperate plant-pollinator network. *PeerJ, 6,* e6025.

34. Halbritter, H., Heigl, H., Auer, W. (2021). *Lotus corniculatus*. In: PalDat - A palynological database. Retrieved from https://www.paldat.org/pub/Lotus_corniculatus/306119; Accessed 23 March 2024.

35. World Flora Online. (2024). *Rosaceae*. Retrieved from http://www.worldfloraonline.org/taxon/wfo-7000000532. Accessed 28 March 2024.

36. The UK Pollen Monitoring Scheme. (2024). *Insect groups visiting PoMS target flowers*. Available at: https://ukpoms.org.uk/flower-charts. Accessed 1 May 2024.

37. Bombosi, P., & Heigl, H. (2020). *Crataegus monogyna*. In: PalDat - A palynological database. Retrieved from https://www.paldat.org/pub/Crataegus_monogyna/304040; Accessed 29 March 2024.

38. Kęsy, M. (2021). The structure of *Prunus avium* L. crops and their importance for pollinating insects in seed orchards in Poland. *Folia Biologica et Oecologica*, 17, 79-83.

39. Forcone, A., Aloisi, P.V., Ruppel, S., & Muñoz, M. (2011). Botanical composition and protein content of pollen collected by *Apis mellifera* L. in the north-west of Santa Cruz (Argentinean Patagonia). *Grana*, 50(1), 30-39.

40. Halbritter, H., Heigl, H. & Auer W. (2021). *Prunus avium*. In: PalDat - A palynological database. Retrieved from https://www.paldat.org/pub/Prunus_avium/306332; Accessed 1 April 2024.

41. Chelifer.com. (n.d.). *Laurel Extra-floral Nectary*. Retrieved April 1, 2024, from https://www.chelifer.com/?page_id=2438. Accessed 1 April 2024.

42. Halbritter, H., Heigl, H. & Auer W. (2021). *Prunus laurocerasus*. In: PalDat - A palynological database. Retrieved from https://www.paldat.org/pub/Prunus_laurocerasus/306335; Accessed 1 April 2024.

43. Elzebroek, T., & Wind, K. (2008). *Guide to Cultivated Plants*. Wallingford: CAB International. ISBN 978-1-845593-356-2.

44. Auer, W. (2021). *Malus domestica*. In: PalDat - A palynological database. Retrieved from https://www.paldat.org/pub/Malus_domestica/306280; Accessed 1 April 2024.

45. Oxford University Herbaria. (n.d.). *Rubus*. Retrieved from https://herbaria.plants.ox.ac.uk/bol/plants400/Profiles/qr/Rubus. Accessed 1 April 2024.

46. Bombosi, P. (2016). *Rubus fructicosus*. In: PalDat - A palynological database. Retrieved from https://www.paldat.org/pub/Rubus_fructicosus/301336; Accessed 1 April 2024.

47. Heigl, H. (2020). *Rubus idaeus*. In: PalDat - A palynological database. Retrieved from https://www.paldat.org/pub/Rubus_idaeus/304009; Accessed 1 April 2024.

48. Flora of China. (n.d.). *Cotoneaster (includes most of the world's Cotoneaster species)*. Retrieved from www.efloras.org. Accessed 1 April 2024.

49. Mountain, M.F. (2021). *Trees and Shrubs Valuable to Bees*. International Bee Research Association. ISBN 978-1-913811-08-2.

50. Heigl, H. (2022). *Cotoneaster dammeri*. In: PalDat - A palynological database. Retrieved from https://www.paldat.org/pub/Cotoneaster_dammeri/306506; Accessed 1 April 2024.

51. Mosseler, A., Major, J., Ostaff, D. & Ascher, J. (2020). Bee foraging preferences on three willow (Salix) species: Effects of species, plant sex, sampling day, and time of day. *Annals of Applied Biology*, 177, 333–345.

52. Woodland Trust. (n.d.). *Goat Willow*. Retrieved from https://www.woodlandtrust.org.uk/trees-woods-and-wildlife/british-trees/a-z-of-british-trees/goat-willow/ Accessed 1 April 2024.

53. Dötterl, S., Glück, U., Jürgens, A., Woodring, J. & Aas, G. (2014, March 27). Floral reward, advertisement and attractiveness to honey bees in dioecious *Salix caprea*. *PLoS One*, 9(3).

54. Hesse, M., Halbritter, H., Heigl, H. & Auer, W. (2021). *Salix caprea*. In: PalDat - A palynological database. Retrieved from https://www.paldat.org/pub/Salix_caprea/306378; Accessed 2 April 2024.

55. Costa, J., Castro, S., Loureiro, J. & Barrett, S.C.H. (2017). Experimental insights on Darwin's cross-promotion hypothesis in tristylous purple loosestrife (*Lythrum salicaria*). *American Journal of Botany, 104* (4), 616–626.

56. Halbritter H., Weber M., Heigl H. & Auer W. (2021). *Lythrum salicaria*. In: PalDat - A palynological database. Retrieved from https://www.paldat.org/pub/Lythrum_salicaria/306279; Accessed 28 April 2023.

57. Alford, D.V. (2016). *Pests of Fruit Crops: A Colour Handbook, Second Edition*. CRC Press. ISBN 978-1-482254-21-1.

58. The Xerces Society. (2016). *Gardening for Butterflies: How You Can Attract and Protect Beautiful, Beneficial Insects*. Timber Press. ISBN 978-1-604695-98-4.

59. Halbritter H. & Heigl H. (2020). *Epilobium angustifolium*. In: PalDat - A palynological database. Retrieved from https://www.paldat.org/pub/Epilobium_angustifolium/304240; Accessed 6 April 2024.

60. Hesse, M. (1981). Pollenkitt and viscin threads: their role in cementing pollen grains. *Grana, 20*(3), 145-152.

61. Porter, D.M. & Sytsma, K.J. (2023). "Sapindaceae". *Encyclopedia Britannica*. Retrieved from https://www.britannica.com/plant/Sapindaceae. Accessed 7 April 2024.

62. Woodland Trust. (n.d.). *Horse Chestnut*. Retrieved from https://www.woodlandtrust.org.uk/trees-woods-and-wildlife/british-trees/a-z-of-british-trees/horse-chestnut/. Accessed 15 April 2024.

References

63. Halbritter, H., Sam S. & Heigl, H. (2020). *Aesculus hippocastanum*. In: PalDat - A palynological database. Retrieved from https://www.paldat.org/pub/Aesculus_hippocastanum/304179; Accessed 7 April 2024.

64. Huxley, A. (Ed.). (1992). *New RHS Dictionary of Gardening*. Macmillan. ISBN 0-333-47494-5.

65. The Canadian Encyclopedia. (n.d.). *Maple*. Retrieved from https://www.thecanadianencyclopedia.ca/en/article/maple. Accessed 7 April 2024.

66. Sam S., Auer W. & Halbritter H. (2020). *Acer platanoides*. In: PalDat - A palynological database. Retrieved from https://www.paldat.org/pub/Acer_platanoides/304175; Accessed 15 April 2024.

67. CAB International. (2022). *Acer pseudoplatanus (sycamore)*. CABI Compendium. doi:10.1079/cabicompendium.2884.

68. Sam S., Auer W. & Halbritter H. (2020). *Acer pseudoplatanus*. In: PalDat - A palynological database. Retrieved from https://www.paldat.org/pub/Acer_pseudoplatanus/304173; Accessed 2 May 2024.

69. Halbritter H., Heigl H., Auer W., Gastaldi C., Geier C., Bouchal J. & Grímsson F. (2024). *Tilia cordata*. In: PalDat - A palynological database. Retrieved from https://www.paldat.org/pub/Tilia_cordata/306671; Accessed 16 April 2024.

70. AgMRC. (2018). *Rapeseed*. Agricultural Marketing Resource Center. Retrieved from https://www.agmrc.org/commodities-products/grains-oilseeds/rapeseed. Accessed 17 April 2024.

71. Stanley, D.A., Gunning, D., & Stout, J.C. (2013). Pollinators and pollination of oilseed rape crops (*Brassica napus* L.) in Ireland: ecological and economic incentives for pollinator conservation. *J Insect Conserv, 17*, 1181–1189.

72. Diethart B. & Heigl H. (2020). *Brassica napus*. In: PalDat - A palynological database. Retrieved from https://www.paldat.org/pub/Brassica_napus/303973; Accessed 18 April 2024.

73. Koutroumpa, K., Theodoridis, S., Warren, *et al*. (2018, December 6). An expanded molecular phylogeny of Plumbaginaceae, with emphasis on Limonium (sea lavenders): Taxonomic implications and biogeographic considerations. *Ecology and Evolution, 8*(24), 12397-12424.

74. Auer, W. (2021). *Limonium vulgare*. In: PalDat - A palynological database. Retrieved from https://www.paldat.org/pub/Limonium_vulgare/306269; Accessed 18 April 2024.

75. Janssens, S.B., Smets, E.F. & Vrijdaghs, A. (2012). Floral development of Hydrocera and Impatiens reveals evolutionary trends in the most early diverged lineages of the Balsaminaceae. *Annals of Botany, 109*(7), 1285-1296.

76. Fair To Nature Farming. (n.d.). *Which Flowers Are the Best Source of Nectar?* Retrieved from https://web.archive.org/web/20191214024659/http://www.conservationgrade.org/2014/10/which_flowers_best_source_nectar/. Accessed 18 April 2024.

77. Halbritter H., Heigl H. & Auer W. (2021). *Impatiens glandulifera*. In: PalDat - A palynological database. Retrieved from https://www.paldat.org/pub/Impatiens_glandulifera/306259; Accessed 18 April 2024.

78. Heigl, H. (2020). *Erica cinerea*. In: PalDat - A palynological database. Retrieved from https://www.paldat.org/pub/Erica_cinerea/304302; Accessed 21 April 2024.

79. Britannica, The Editors of Encyclopaedia. (2020, April 1). *Heath. Encyclopedia Britannica*. Retrieved from https://www.britannica.com/plant/heath. Accessed 22 April 2024

80. Halbritter H. & Heigl H. (2020). *Erica carnea*. In: PalDat - A palynological database. Retrieved from https://www.paldat.org/pub/Erica_carnea/304303. Accessed 22 April 2024.

81. Royal Botanic Gardens, Kew. *Calluna vulgaris (L.) Hull*. Retrieved 22 April 2024.

82. Koch, H., Woodward, J., Langat, M., *et al*. (2019). Flagellum Removal by a Nectar Metabolite Inhibits Infectivity of a Bumblebee Parasite. *Current Biology, 29*, 3494–3500.

83. Goulson, D. & Wright, N.P. (1998). Flower constancy in the hoverflies *Episyrphus balteatus* (Degeer) and *Syrphus ribesii* (L.) (Syrphidae). *Behavioral Ecology, 9*(3), 213-219.

84. Stout, J.C. & Parnell, J.A. (2007). Pollination ecology and seed production of *Calluna vulgaris* (heather) on heathland and moorland in Co. Wicklow, Ireland. *Irish Naturalists' Journal, 28*(2), 63-72.

85. Halbritter H. & Heigl H. (2020). *Calluna vulgaris*. In: PalDat - A palynological database. Retrieved from https://www.paldat.org/pub/Calluna_vulgaris/304299; Accessed 22 April 2024.

86. Coats, A.M. (1992). Garden Shrubs and Their Histories (1964). s.v. "*Rhododendron*". Retrieved from http://www.countrysideinfo.co.uk/rhododen.htm#Introduction%20to%20Britain. Accessed 22 April 2024.

87. Cross, J.R. (1975). *Rhododendron ponticum* L. *Journal of Ecology, 63*(1), 345–364.

88. Stout, J.C., Parnell, J.A.N., Arroyo, J., *et al*. (2006). Pollination Ecology and Seed Production of *Rhododendron ponticum* in Native and Exotic Habitats. *Biodivers Conserv, 15*, 755–777.

89. Stout, J.C. (2007). Pollination of invasive *Rhododendron ponticum* (Ericaceae) in Ireland. *Apidologie, 38*(2), 198-206.

90. Montaner, C., Floris, E. & Alvarez, J.M. (February 2001). Geitonogamy: a mechanism responsible for high selfing rates in borage (*Borago officinalis* L.). *Theoretical and Applied Genetics, 102*(2–3), 375–378.

91. Halbritter, H. & Heigl, H. (2020). *Borago officinalis*. In: PalDat - A palynological database. Retrieved from https://www.paldat.org/pub/Borago_officinalis/304382; Accessed 23 April 2024.

92. Weryszko-Chmielewska, E. (2003). Morphology and anatomy of floral nectary and corolla outgrowths of *Myosotis sylvatica* Hoffm. (Boraginaceae). *Acta Biologica Cracoviensia Series Botanica, 45*, 43-48.

References

93. Halbritter H., & Heigl H. (2020). *Myosotis arvensis*. In: PalDat - A palynological database. Retrieved from https://www.paldat.org/pub/Myosotis_arvensis/304389; Accessed 23 April 2024.

94. Graves, M., Mangold, J. & Jacobs J. (2018). *Biology, Ecology and Management of Blueweed*. Montana State University. Accessed 23 April 2024.

95. Klinkhamer, P.G.L. & de Jong, T.J. (1990). Effects of Plant Size, Plant Density and Sex Differential Nectar Reward on Pollinator Visitation in the Protandrous *Echium vulgare* (Boraginaceae). *Oikos, 57*(3), 399–405.

96. Loos, C., Halbritter, H., Heigl, H. (2020). *Echium vulgare*. In: PalDat - A palynological database. Retrieved from https://www.paldat.org/pub/Echium_vulgare/304395; Accessed 23 April 2024.

97. Hickman, J.M., Wratten, S.D. (1996, August 1). Use of *Phacelia tanacetifolia* Strips To Enhance Biological Control of Aphids by Overfly Larvae in Cereal Fields. *Journal of Economic Entomology, 89*(4), 832–840.

98. Halbritter H. & Kratschmer S. (2016). *Phacelia tanacetifolia*. In: PalDat - A palynological database. Retrieved from https://www.paldat.org/pub/Phacelia_tanacetifolia/300737; Accessed 23 April 2024.

99. Halbritter H., Heigl H. & Auer W. (2021). *Ligustrum vulgare*. In: PalDat - A palynological database. Retrieved from https://www.paldat.org/pub/Ligustrum_vulgare/306267; Accessed 23 April 2024.

100. Wiese, N., Fischer, J., Heidler, J., *et al.* (2018). The terpenes of leaves, pollen, and nectar of thyme (*Thymus vulgaris*) inhibit growth of bee disease-associated microbes. *Sci Rep, 8*, 14634.

101. Auer W. (2020). *Thymus vulgaris*. In: PalDat - A palynological database. Retrieved from https://www.paldat.org/pub/Thymus_vulgaris/304449; Accessed 24 April 2024.

102. Ulrich S. (2016). *Mentha aquatica*. In: PalDat - A palynological database. Retrieved from https://www.paldat.org/pub/Mentha_aquatica/300512; Accessed 24 April 2024.

103. Britannica, The Editors of Encyclopaedia. "*lavender*". *Encyclopedia Britannica*, 9 Apr. 2024, https://www.britannica.com/plant/lavender; Accessed 24 April 2024.

104. Littlewood, V. (n.d.). Pollen: Beautiful colours, fascinating form. *Pencil and Leaf*. Retrieved from https://pencilandleaf.valerielittlewood.uk/2010/10/pollenbeautiful-colours-fascinating.html; Accessed 1 May 2024.

105. Halbritter, H., Weber, M. & Heigl, H. (2020). *Lavandula angustifolia*. In: PalDat - A palynological database. Retrieved from https://www.paldat.org/pub/Lavandula_angustifolia/304412; Accessed 24 April 2024.

106. Britannica, The Editors of Encyclopaedia. "*rosemary*". *Encyclopedia Britannica*, 4 Apr. 2024, https://www.britannica.com/plant/rosemary. Accessed 24 April 2024.

107. Dirr, M.A. (2011). *Dirr's Encyclopedia of Trees and Shrubs*. Timber Press. ISBN 978-0881929010

108. Richmond Beekeepers Association. (n.d.). *Pollen Colour Chart*. Retrieved from www.richmondbeekeepers.co.uk/tools-links/pollen-colour-chart.pdf. Accessed 24 April 2024.

109. Halbritter H., Hesse M., Heigl H. & Svojtka N. (2020). *Rosmarinus officinalis*. In: PalDat - A palynological database. Retrieved from https://www.paldat.org/pub/Rosmarinus_officinalis/304409; Accessed 5 May 2024.

110. Herrera, C. & Bazaga, P. (2012). Epigenetic correlates of plant phenotypic plasticity: DNA methylation differs between prickly and non prickly leaves in heterophyllous *Ilex aquifolium* (Aquifoliaceae) trees. *Botanical Journal of the Linnean Society*, Wiley.

111. Invasive Species Council of British Columbia (ISCBC). *English Holly*. Retrieved from https://bcinvasives.ca/invasives/english-holly/. Accessed 24 April 2024.

112. California Invasive Plant Council (Cal-IPC). *Ilex aquifolium*. Retrieved from https://www.cal-ipc.org/plants/profile/ilex-aquifolium-profile/. Accessed 24 April 2024.

113. Woodland Trust. *Holly*. Retrieved from https://www.woodlandtrust.org.uk/trees-woods-and-wildlife/british-trees/a-z-of-british-trees/holly/. Accessed 24 April 2024.

114. Halbritter H., Heigl H., Auer W. & Ulrich S. (2022). *Ilex aquifolium*. In: PalDat - A palynological database. Retrieved from https://www.paldat.org/pub/Ilex_aquifolium/306501; Accessed 24 April 2024.

115. Standifer, L.N. (1966). Fatty Acids in Dandelion Pollen Gathered by Honey Bees, *Apis mellifera* (Hymenoptera: Apidae). *Annals of the Entomological Society of America*, Volume 59, Issue 5, 1005–1008.

116. Bombosi P. & Heigl H. (2021). *Taraxacum officinale*. In: PalDat - A palynological database. Retrieved from https://www.paldat.org/pub/Taraxacum_officinale/305331; Accessed 25 April 2024.

117. Bryant, V. (2020). Why Honey Pollen is Difficult to Interpret. *Bee Culture*, June 1, 2020.

118. Karacetin, E. (2007). *Biotic Barriers to Colonizing New Hosts by the Cinnabar Month Tyria jacobaeae (L.) (Lepidoptera: Arcitiidae)*. Dissertation; Oregon State University.

119. Kirk, W.D.J. (2006). *A Colour Guide to the Pollen Loads of the Honey Bee*. International Bee Research Association. ISBN 0-86098-248-3.

120. Halbritter, H. (2016). *Senecio jacobaea*. In: PalDat - A palynological database. Retrieved from https://www.paldat.org/pub/Senecio_jacobaea/301543; Accessed 25 April 2024.

121. Susic M.C & Farina W.M. (2016). Honeybee floral constancy and pollination efficiency in sunflower (*Helianthus annuus*) crops for hybrid seed production. *Apidologie*, 47, 161–170.

122. Halbritter H., Heigl H. & Svojtka N. (2020). *Helianthus annuus*. In: PalDat - A palynological database. Retrieved from https://www.paldat.org/pub/Helianthus_annuus/304619; Accessed 25 April 2024.

123. Bartlett, J. (2020). *The Pollen Landscape*. Northern Bee Books. ISBN 978-1-914934-06-3.

References

124. Galloway Wild Foods. (n.d.). Sea Aster: Identification, Edibility, Distribution. Retrieved April 27, 2024, from https://gallowaywildfoods.com/sea-aster-identification-edibility-distribution/; Accessed 25 April 2024.

125. Auer, W. (2021). *Aster tripolium*. In PalDat - A palynological database. Retrieved from https://www.paldat.org/pub/Aster_tripolium/305661; Accessed 27 April 2024.

126. Halbritter, H., Heigl, H. & Svojtka, M. (2020). *Tripolium pannonicum*. In PalDat - A palynological database. Retrieved from https://www.paldat.org/pub/Tripolium_pannonicum/304606; Accessed 27 April 2024.

127. Natural History Society of Northumbria. (2020, September 24). A focus on Ivy. From https://www.nhsn.org.uk/ivy-an-autumnal-magnet-for-pollinators; Accessed 27 April 2024.

128. Halbritter, H. & Heigl, H. (2020). *Hedera helix*. In PalDat - A palynological database. Retrieved from https://www.paldat.org/pub/Hedera_helix/304248; Accessed 27 April 2024.

129. Britannica, The Editors of Encyclopaedia. (2024, February 9). *Apiaceae. Encyclopedia Britannica.* https://www.britannica.com/plant/Apiaceae; Accessed 27 April 2024.

130. Pollen Atlas. (n.d.). *Daucus carota*. Retrieved from https://pollenatlas.net/apiaceae/daucus/daucus-carota; Accessed 27 April 2024.

131. Halbritter, H., Heigl, H. & Auer, W. (2021). *Daucus carota*. In PalDat - A palynological database. https://www.paldat.org/pub/Daucus_carota/305987; Accessed 27 April 2024.

132. USDA, Agricultural Research Service, National Plant Germplasm System. (2024). *Daucus carota*. Germplasm Resources Information Network (GRIN Taxonomy). National Germplasm Resources Laboratory, Beltsville, Maryland. URL: https://npgsweb.ars-grin.gov/gringlobal/taxon/taxonomydetail?id=12265. Accessed 28 April 2024.

133. Heigl, H. (2021). *Crocus purpureus*. In: PalDat - A palynological database. Retrieved from https://www.paldat.org/pub/Crocus_purpureus/305006. Accessed 28 April 2024.

134. Rabinowitch, H.D. & Thomas, B. (2023). Chives (*Allium schoenoprasum*). In Edible *Alliums—Botany, Production and Uses*. CAB International. ISBN 978-1-789249-996.

135. Agricultural Research Service, National Plant Germplasm System. (2024). Germplasm Resources Information Network (GRIN Taxonomy). National Germplasm Resources Laboratory, Beltsville, Maryland. URL: https://npgsweb.ars-grin.gov/gringlobal/taxon/taxonomydetail?id=2369. Accessed 29 April 2024.

136. Landry, J-F. (2007). Taxonomic review of the leek moth genus *Acrolepiopsis* (Lepidoptera: Acrolepiidae) in North America. *The Canadian Entomologist*, 139 (3), 319–353.

137. Heigl, H. & Auer, W. (2021). *Allium schoenoprasum*. In PalDat - A palynological database. Retrieved from https://www.paldat.org/pub/Allium_schoenoprasum/306064. Accessed 29 April 2024.

138. Halbritter, H., Heigl, H. & Schneider, H. (2020). *Galanthus nivalis*. In PalDat - A palynological database. Retrieved from https://www.paldat.org/pub/Galanthus_nivalis/304774. Accessed 29 April 2024.

139. Halbritter, H. (2015). *Hyacinthoides hispanica*. In: PalDat - A palynological database. Retrieved from https://www.paldat.org/pub/Hyacinthoides_hispanica/300049. Accessed 30 April 2024.

140. Halbritter, H., Aktuna, G., Heigl, H. & Svojtka, M. (2020). *Asparagus officinalis*. In: PalDat - A palynological database. Retrieved from https://www.paldat.org/pub/Asparagus_officinalis/304760. Accessed 1 May 2024.

www.ingramcontent.com/pod-product-compliance
Lightning Source LLC
Chambersburg PA
CBHW041107280326
41928CB00062B/3450